Geoinformatics, Precision Farming and Nanotechnology

NIPA® GENX ELECTRONIC RESOURCES & SOLUTIONS P. LTD.
New Delhi-110 034

About the Authors

Dr Manpreet Kour is Doctorate on simulation modeling in Agronomy, presently working as Associate Professor AICRP on Weed Management, Division of Agronomy, SKUAST Jammu, with a teaching, research, and extension experience of about 17 years. She has teaching experience of more than 150 credit hours and handled seven research projects. Her publications include 33 research papers, 50 abstracts, and 68 extension publications along with eight manuals, eight popular articles, 37 chapters in book/compendium, two research bulletins, and presented more than 22 papers in various seminars and symposia. Apart from these 10 recommendations and technologies have been developed. Her contribution has been recognized by 16 awards/appreciations.

Dr Vikas Sharma is Professor and Head, Division of Soil Science and Agricultural Chemistry at the Sher-e-Kashmir University of Agricultural Sciences and Technology of Jammu, India. He obtained his doctorate from the Punjab Agricultural University, Ludhiana, India and did postdoc under the Raman Fellowship from West Virginia University, USA. He has been a Visiting Professor at the Auburn University, USA. He is the founder Head of Centre for Organic and Natural Farming and has been instrumental in establishing the center from scratch. His research interests include digital soil mapping, pedometrics, soil carbon, soil conservation, and land degradation. Most of his work pertains to soils in Indian Himalayas. He has worked on soil erosion risk mitigation, carbon pools, and carbon losses from the Himalayan soils and nutrient dynamics in the erosion-prone foothill of Himalayas. He has utilized his expertise in remote-sensing and GIS in assessing and mapping soil resources in the region. He is presently focusing on utilizing machine learning models in soil mapping. He is also involved in teaching and advising students' research. He has executed multiple projects and presently is the PI of the prestigious NAHEP project at SKUAST Jammu. He has over 90 research articles and 30 book chapters to his credit. His latest authored book is on Soils in the Hindukush Himalayas has been published by Springer.

Dr Harsimran Singh Bindra is Doctorate in nanoscience, presently working as Assistant Professor (Contractual) Division of Biotechnology, SKUAST Jammu and have teaching, research and extension experience of about 8 years. His publications include 12 research papers, eight popular articles, four chapters in book/compendium, and presented more than four papers in various seminars and symposia. Apart from these he owns one patent.

Geoinformatics, Precision Farming and Nanotechnology

Manpreet Kour
Associate Professor
AICRP on Weed Management
Division of Agronomy
Sher-e-Kashmir University of Agricultural Sciences & Technology
Jammu, Jammu and Kashmir, India

Vikas Sharma
Professor and Head
Division of Soil Science and Agricultural Chemistry
Sher-e-Kashmir University of Agricultural Sciences and Technology
Jammu, Jammu and Kashmir, India

Harsimran Singh Bindra
Assistant Professor (Contractual)
Division of Biotechnology
Sher-e-Kashmir University of Agricultural Sciences & Technology
Jammu, Jammu and Kashmir, India

NIPA® GENX ELECTRONIC RESOURCES & SOLUTIONS P. LTD.
New Delhi-110 034

NIPA® GENX ELECTRONIC RESOURCES & SOLUTIONS P. LTD.

101,103, Vikas Surya Plaza, CU Block
L.S.C.Market, Pitam Pura, New Delhi-110 034
Ph : +91 11 27341616, 27341717, 27341718
E-mail:newindiapublishingagency@gmail.com
www: www.nipabooks.com

For customer assistance, please contact
Phone: + 91-11-27 34 17 17
Fax: + 91-11- 27 34 16 16

Print ISBN: 978-93-58877-91-5

ebook ISBN: 978-93-58877-25-0

Composed and Designed by NIPA

Preface

India's agrifood industry will have a difficult task in 2050 to feed nearly 1.7 billion people on a limited amount of arable land, water, and energy. At 140 million hectares, our net sown area has all but stagnated, and there is little chance that it will rise. The process of intensifying agriculture has resulted in the degradation of natural resources, specifically soil and water. Our current input application strategy ignores intrafield variability and is based on standard recommendations for a composite sample or visual symptoms of crops. Since the mean values used to make recommendations are rarely observed in a particular field, it indicates either excessive or suboptimal application of inputs. In addition to decreasing input use efficiency, it degrades the environment. As a result, India is dedicated to bringing about a technologically advanced and demand-driven revolution to address the issues of increased food demand better farmer livelihood opportunities, and sustainable farming for increased agricultural growth.

Geoinformatics, Precision Farming, and Nanotechnology are climate-smart alternatives to sustainable agriculture, under dwindling resources having modern tools and techniques. These are launched to optimize agriculture production, save on input cost, and minimize adverse impact on environment. They advocate the application of right inputs in right amounts at right time and at right place using the right technologies or techniques taking soil, crop variability into consideration. Information generation and collection were accelerated by the development of spectral and hyperspectral sensors as well as high-resolution images obtained by satellites along with innovations in nanotechnology, such as nanoscale biosensors and nanoparticles, that appeared as promising agents for plant growth, in use of fertilizers, pesticides and thus in agriculture productivity

The Authors express optimism that the student body will utilize this book to gain insight into the fields of precision, nanotechnology, and geoinformatics.

Authors extend their sincere thanks and indebtedness to the distinguished authors of various standard textbooks and other publications or sources, which have benefitted in the research of this book. In several cases, it has not been possible to obtain permission for reproduction, for which authors and publishers offer their apologies. If any errors may kindly be pardoned and brought to the notice for further improvement.

Deeply felt gratitude is due to the Hon'ble Vice Chancellor SKUAST Jammu for inspiration and guidance. Sincere thanks are also due to colleagues and friends for their constant help and encouragement.

We hope this present publication will be useful to students and academicians working in agriculture and allied sciences.

Authors

Contents

1

Geographical Information System

Abstract

A geographical information system (GIS) is a powerful tool that enables computer-based manipulation of geographical information. It has various hardware and software components that allow data input, storage, management, and reporting. A GIS database consists of different types of data formats, such as vector and raster, that define the objects, relationships, and geometric features of the spatial information. Furthermore, some examples of GIS applications are provided to demonstrate how GIS data can be used for different purposes, viz., crop and yield estimation, soil mapping, flood impact, and drought estimation.

Keywords: Geographical information system, raster, vector, GIS data management, data elevation system

Introduction

A geographical information system (GIS) is a kind of database that has software tools for handling, analyzing, and presenting spatial data and descriptions of phenomena that have a location component. This system can also involve human users and support staff, procedures and workflows, a body of knowledge about relevant concepts and methods, and institutional organizations in a wider sense. GIS technology is a key part of spatial data infrastructure, which the White House describes as "the policies, technologies, standards, human resources, and related activities that are necessary to acquire, process, distribute, use, maintain, and preserve spatial data." GIS is a technology that enables finding meaningful patterns and insights from data. It is changing rapidly and offering a new framework and processes for understanding (ESRI 2022). GIS can be a useful technology for collecting and using geographic information to support the development of agriculture. For example, a digital map has more value than the same map printed on paper because the digital version can be integrated with other sources of data or information based on graphical presentation. The GIS software allows synthesizing large amounts of different data, combining different layers of information to manage and retrieve the data in a more useful way. Therefore, GIS provides a powerful tool for agricultural scientists to better serve the farmers and farming community by answering their questions and helping them in better decision-making to implement planning activities for the development of agriculture (ESRI 2022).

Goals and objective of GIS

1. Maximize the effectiveness of higher cognitive processes and design
2. Provide cost-effective recommendations for the distribution and handling of information
3. Eliminate redundant databases by reducing duplications
4. Integrate information from multiple sources
5. Get new information from complex analysis and queries that involve geographical reference information
6. Quickly and cost-effectively update information
7. Understand soil information and determine where crops should be planted and how to care for soil nutrition
8. Increase productivity and profitability through more efficient natural resource management and reduce the risk of marginalization and vulnerability of small and marginalized farmers

Types

A GIS is a system of computer programs that can be used to show, manipulate, and analyze spatial data from different sources. The data sets used in a GIS are usually divided into different categories for easier storage and use. Each data set that a GIS can handle has two main parts: graphical (spatial) data and tabular (feature) data. Spatial data is data that has a location and is displayed graphically on the computer screen. Each piece of graphical data is called a feature. Features can be points, lines, or polygons. The feature or tabular data is text or numerical data that describes each of the features. The tabular data is linked to the graphical data and has a unique ID number for each point, line, or polygon. Examples of tabular data can include things such as addresses, coordinates, area, length, sales data, and street names. There are many possibilities for data association between tabular and graphical data.

Spatial information can frequently be similarly divided into most important sorts.

- The first category is raster type data, which is generally included inside the shape of images, like aerial photographs or imported scans of old maps. The location and color production cost of each pixel that makes up the image are stored in the raster information
- The second type of information is entirely vector-based. An amalgam of line work, polygons, and curvilinear data may constitute a vector. This statistic is stored via an amalgamation of locally distinctive points, lines, and arcs

Raster images can lose quality and become pixelated while scaling. On the other hand, vector information can be scaled to any size without losing any quality. Another way to look at information sorts is to know that some information is discrete while others are continuous.

- Discrete information, on the other hand, is usually mostly vector-based and contains unique statistics on unique factors with ranges in between
- On the other hand, continuous information is usually completely raster-based and there are no gaps. Statistics can be found anywhere within the range of a raster

When each piece of data is loaded and shown inside the GIS mapping window, it is called a layer. But when a few layers are used inside the same map window, they can be stacked, color-coded and symbolized to represent a myriad of map compositions. Stacked data sets can also be manipulated by changing the color ramps and hatching and shading and/or the transparency degrees to reveal new relationships that previously could not be seen in conventional maps. The main purpose of GIS is the ability to describe, analyze, and display large amounts of spatial data in a way that a simple virtual map cannot.

History

Early History

The field of GIS began in the 1960s with the development of computers and the early development of quantitative and computational earth science. The early work of GIS included the need for analysis by the education community. In 1988, efforts such as the foundation of National Center for Geographic Information and Analysis and the development of the light-emitting diode (LiDiode) by Michael Goodchild, which focused on uncertainty in geographical information, the development of discrete global grids and the formalization of analysis on key geospatial information science topics, such as spatial analysis and visual images, fueled a quantitative revolution in the world of geology and laid the foundations for GIS (ESRI 2022).

- *First GIS*: Roger Tomlinson founded, designed, and implemented the first Canadian GIS in 1963. In the same year, Tomlinson developed the world's first computerized GIS. The Canadian government hired Tomlinson in 1963 to compile a comprehensive inventory of Canada's natural resources. Tomlinson's vision was to bring together natural resource data from all the provinces using computers (ESRI 2022).
- *Harvard laboratory*: In 1964, Howard Fisher was working at North West University when he created SYMAP, which was one of the first ever computer mapping software programs. He then found the Harvard Laboratory for Computer Graphics in 1965, and it quickly became a hub for research in spatial analysis and visualization, as well as the source of some of the first computer map-making programs. A bunch of different people from different backgrounds came together to create GIS, and a lot of the ideas behind it came from the laboratory (ESRI 2022).
- *Environmental Systems Research Institute (ESRI)*: In 1969, Jack Dangermond, a member of the Harvard Laboratory, and his wife Laura started ESRI. They wanted to help land resource managers and land use planners, so they used computer maps and spatial analysis. This showed that GIS could be used to solve problems. Since then, ESRI has created a lot of the modern GIS mapping and analysis tools that people are interested in, like their software and GIS-style workflows (ESRI 2022).
- *GIS commercialization*: As computers got more powerful, ESRI came up with powerful GIS tools and techniques that people could use. They worked on projects that addressed real-world problems, which improved their software tools and eventually led to GIS being commercially available. ESRI's work in spatial analysis and planning has also become well-known in the academic world. In 1981, they released the first commercial version of GIS, ARC/INFO, to help improve analysis. This was the start of ESRI's transition into a software company (ESRI, 2022).
- *GIS update today*: GIS is a tool that helps people create their own digital maps that can help them solve real-world problems. It is a vision that is coming to life quickly—a global GIS database that covers all kinds of topics, and it is inspired by GIS's growth as a way to share and collaborate data. Hundreds of thousands of companies are sharing their

work and creating billions of maps each day to tell stories and uncover trends, patterns, and connections around the world (ESRI 2022).

Components of a Geographic Information System

Geographic information system information hubs are typically composed of a combination of software programs, hardware components, and statistics, with all customers to the GIS being connected to the same statistical units. This ensures uniformity of data and will help to avoid duplication of effort. Furthermore, due to the web-first approach, even remote customers can access the same data as everyone else (Sarma 2021). There are five components of GIS, which are described in the following text.

Hardware

Hardware is a collection of devices, such as computers, smartphones, tablets, laptops, monitors, printers, and scanners, that can store large geographical information sets, GIS software, and applications. Additionally, there are many server-based computers that can be connected to a server and run GIS software either on a network machine or in a cloud. The ArcGIS server is a software that makes geographic information accessible to those with an internet connection. Hardware is composed of a motherboard, a disk drive, a processor, a RAM, an output device, and a disk. The RAM is a device that stores data and allows for the temporary loading of programs. The disk is an external storage medium, such as a CD or USB drive, and the output device is a monitor with a variety of CRT, LCD, and LED displays.

Software

The GIS software is a component of GIS programming that provides devices with the ability to access and modify spatial data. It facilitates the process of querying, editing, displaying, and running GIS data. This software is based on the RDBMS, which stores the information in query Tools, such as SQL, for querying, inserting, deleting, and other operations that are compatible with the database management system. GUI is a graphical user interface that facilitates a smooth interaction between the software and the user. Layout is a useful window for creating a map. Some of the popular GIS software programs include ArcGIS, ArcView, QGIS, and Adventure GIS.

Data

Geographic Information System data is the most significant and costly component of the system. GIS data is a combination of tabular and graphical data, which can be raster or vector graphics. Additionally, GIS software can either be used to generate either type of data or can be purchased.

The digitization process involves the digitization of analogue data or paper formats, and the registration of a raster image with a few GCPs (ground control points) or known coordinates. This process is commonly referred to as rubber sheeting and georeferencing. Upon digitization, polygons, lines, and points are formed on the raster image, and the raster picture itself is enrolled with arranges, commonly referred to as amending the image. Most of the registered images are exported using TIFF format. GIS data can be either raster or vector, as previously stated.

- Raster images contain data that is stored in a cellular format. These images may be satellite images, aerial photographs, or Digital Elevation Models (DEMs). Generally, raster images contain continuous data.
- Distinctive vector data is a type of data structure that stores data using the x and y coordinate system. There are three main types of vector data: area, points, and lines.

People

Individuals are the primary users of Geographic Data Framework (GDF). The GDF software is managed by them. Recent advances in hardware and software have made the GDF software user friendly. As a result, people are increasingly familiar with and utilizing computers for GDF tasks due to their cost-effectiveness. These tasks may include creating basic guides or conducting more advanced GIS research. Therefore, the most essential element of a successful GDF is its personnel. Individuals are client of GDF. The GIS software is run by them.

Methods

For GIS activities to be successful, it is essential to have a comprehensive plan and rules for business activity. Strategies may vary depending on the organization's associations. Every organization has a documented GIS operational process plan. These documents address the following questions related to GIS: number of GIS experts required; software and hardware required for GIS; storage of data; type of DBMS to use; and more. A comprehensive plan will provide answers to all of these queries.

Methodology

The GIS functionalities can be employed to generate an analysis model to address the issue, determine the goals of the project, generate a project database that contains the necessary information to address the issue, and present the results of the analysis in a standard GIS analysis process. The procedure in GIS involves various steps:

Step 1: The objective of the analysis must be identified in the first step of the procedure. As it solved many queries like:

- What issue must be resolved? How is it now resolved? Is there different ways to use GIS to solve the problems?
- What are the project's final products, including working maps, maps that are ready for presentation, and reports?
- Who are these products aimed at—the general public, technicians, planners, and government officials?
- Will other uses be made of the data? What requirements must be met for these?
- This step is important because the answers to these questions determine the project's scope and then how the analysis are used into practice.

Step 2: Make an undertaking of information base

The next step in the process of creating an undertaking information base is to create a project database. This project database is divided into three stages: database design, automated data collection and collection, and database administration. The selection of the coordinate system to be used, the establishment of the study area boundaries, and the specification of the desired feature attributes are all included in the database design. The spatial data should be determined in accordance with the objectives of the analysis. Additionally, digitization or conversion of data from other systems or formats into a suitable format is a process that can be automated. Verification of the data and correction of any errors is also part of database management. The creation of the project database is a critical and time-consuming part of the project. The completeness of the data is essential for the accuracy of the analysis's findings.

Step 3: Break down of the information

The third step is to break down the information. Data analysis of GIS can range from making straightforward maps to creating intricate spatial models. A model is a representing the real-world and further used to simulate the process, forecast a result, or examine a problem. Applying one to three categories of

GIS functionality to some spatial data is the basis of a spatial model. The functions are as follows:

- Functions for geometric modeling include calculating distances, generating buffers, and calculating areas and perimeters.
- Functions for coincidence modeling include overlaying datasets to locate locations where values coincide.
- Functions for adjacency modeling include allocating, pathfinding, and redistricting. Using a GIS makes it possible to quickly carry out analyses that would be either impossible or extremely time-consuming to carry out manually. By rerunning the analysis and changing your methods or parameters, you create alternative scenarios.

Step 4: Present the results with efficiency

The results of a GIS analysis should be presented in an efficient manner. Generally, the most effective way to present the findings is on a map. Additionally, charts and reports of the selected data can be printed individually, incorporated into other applications, or placed on the map.

Tools in GIS

Tools and strategies: GIS mapping tools can be divided into four main categories: desktop, web, server, and specialized. With the widespread availability of satellite and aerial imagery, digitization is becoming the primary method of extracting geographic facts. This digitization involves the immediate tracing of the geographical shape on a separate digitizing tablet, as opposed to the traditional method of tracing the shape on a satellite or aerial image. The key concepts and techniques employed in GIS spatial relationships and spatial facts are virtual facts, use and access of existing facts, mixture of facts, modeling strategies, and map algebra. Additionally, spatial records and spatial analysis are also being developed. Technology used in GIS data introduction is based on virtual statistics, with various digitized facts creation strategies. GIS statistics can be used to analyze a wide range of unique data sources in a variety of specialized formats. The primary requirement for supply facts is an understanding of the locations of the variables. This can be done by annotating the location with the use of latitude, longitude, and elevation coordinates, or by utilizing geocoding structures such as postal codes. Additionally, any variable that can be spatially located can be directly inputted into the GIS. Various types of facts can be entered into the GIS in a map-like format. GIS facts can be used to represent actual global items, such as roads and land use as well as virtual facts. These real global items can be divided into discrete items, such as a house, and nonstable fields, such as rainfall quantities or elevation.

Raster

A raster data kind is, in essence, any sort of virtual photograph. Anyone who's acquainted with virtual pictures will apprehend the pixel because the smallest character unit of a photograph. An aggregate of those pixels will create a photograph, awesome from the usually used scalable vector pics, which can be the premise of the vector version. While a virtual photograph is involved with the output as illustration of reality, in an image or artwork transferred to computer, the raster data kind will replicate an abstraction of reality. Aerial snap shots are one usually used shape of raster facts, with handiest one purpose, to show an in-depth photograph on a map or for the functions of digitization. Other raster facts units will contain statistics regarding elevation, a DEM (virtual elevation model), or reflectance of a selected wavelength of light.

Vector

It is possible to display vector facts as vector images used on traditional maps, while raster facts will appear as a photograph which may have a blocky appearance for item boundaries. Additionally, vector facts may be easier to register, expand, and repurpose. This can make it simpler to combine vector layers from distinct sources. Furthermore, vector facts are particularly well suited to a relational database environment.

Advantages and Disadvantages of Raster and Vector

There are benefits and disadvantages to the usage of a raster or vector statistics model to symbolize reality. Raster statistics sets record a fee for all factors in the area covered which may also require greater storage area than representing statistics in a vector format that can keep statistics simplest wherein needed. Raster statistics additionally lets in smooth implementation of overlay operations, which can be greater hard with vector statistics. Vector statistics can be displayed as vector portraits used on conventional maps, while raster statistics will appear as a photo which could have a blocky look for item boundaries. Vector statistics can be less complicated to register, scale, and reproject. This can simplify combining vector layers from one-of-a-kind sources. Vector statistics are highly compatible with relational database environment (ESRI 2016).

Application of GIS

- In school geography to enriched current educational environment can depict erosion risk analysis, suitability of land in the construction of the dam, etc. (Wiegand 2001)

- Environmental Impact Assessment (EIA) is a methodical procedure that examines the environmental effects of proposed projects. The use of GIS, which are extremely effective tools for impact assessment, is likely to rise significantly in the near future (Şahin and Kurum 2002).
- Utilization of GIS in hydraulic modelling in conjunction to manage a water distribution network, an arrangement of GIS and modeling enables network administrators to better analyze malfunctions, resulting in a response that is quicker (Abdelbaki et al., 2017).
- In medical science in early diagnose of the diseases, e.g., cancer by assessing connections between the conveyance of malignant growth cases and geo-ecological elements is huge in deciding the reasons for disease. Further a cancer density map was created and correlations between cancer types and geographical factors were examined in order to determine the cancer occurrence density by using local census survey (Yomralioglu et al., 2009)
- Application in agriculture is discussed in detail in the following text.

Advantages of GIS in Agriculture

Weather risk management services uses remote-sensing and GIS technology for actual-time evaluation and forecasts of climate, floods and tracking crop location sown, lost or damaged because of floods and rainfall. Agriculture is one of the principal sectors in Indian economy. India being a world's biggest manufacturer of rice, wheat and pulses, ought to maintain up its tempo with developing virtual technology. Over the beyond few decades, remote-sensing and GIS have grown exponentially in lots of sectors for visualization, tracking, control, and capability development. Remote-sensing and GIS generation allow businesses to get dependable records of natural and man-made features, or processed and interpreted appropriately phenomenon happening over the earth's surface without making any physical contact. Nowadays, farmers rely upon those technological improvements. Brief descriptions of remote-sensing services supplied through Weather Risk Management Services (WRMS) are given below.

Crop Estimation

The GIS biomass and yield of crop can be precisely and consistently predicted through crop estimation. It quickly provides the knowledge need to make important choices regarding the harvesting of crop and proper handling of post-harvest techniques. Further it assists in optimum allocating of the resources (labor, machinery, etc.) in accordance with yield forecasts and managing their logistics (opening dates, inputs, harvesting machine distribution, etc.). It can further helpful for the mapping and monitoring of numerous sown location estimates, which are provided by satellite statistics such as Sentinel-1 and Sentinel-2. Additionally it also allows for the pinpointing of specific stage of sown location, which is essential for the assessment of crop loss due to the prevalence of catastrophic events.

Normalized Difference Vegetative Index (NDVI)

Geographical information system techniques are frequently employed for assessing natural resources, determining the land changes, and performing related planning tasks. The methodology is based on correlation with remotely sensed vegetation data in the form of the NDVI. This index's primary use is to track the amount of vegetative cover. NDVI is a widely used tool for assessing plant dynamics, particularly to determine crop fitness. It provides an opportunity to gain a deeper understanding of crop phenology, as it elucidates the chronology of crops and their correlation to climate and weather conditions. NDVI is quantified by the use of mathematical calculations of spectral bands within the satellite photograph, which is used to measure future plant health, as it has a strong relationship with inexperienced biomass, suggesting a healthy plant or crop.

Crop Diseases Identification

By utilizing a GIS to organize epidemiological data in the same manner as other farm data, plant disease management techniques can be enhanced. Selecting the appropriate scale and the types of data to include at each scale is a skill in developing GIS databases. The provision of GIS will enable farmers, extension agents, pest control consultants, and other stakeholders to assess plant disease issues within a spatial framework. The GIS technology provides a cost-effective solution for the spatial distribution of disease and pest records over a large geographical area. With the use of satellite images and spatial evaluation methods, areas that have been identified as infested with pests such as the Mealybug Bug, the Plant Hopper Bug, and the White Fly are analyzed to provide a comparison of the infested regions.

Soil Properties

Mapping the geographical variability of soil parameters is essential for modeling agricultural productivity, food safety, and the environment. Soil is a fundamental factor in agricultural control practices, as it has a direct impact on yield output. Surface maps of soil attributes were created using GIS techniques, and sampling-broad analysis using Global Positioning System (GPS) was found to be a potentially useful tool. Changes in agricultural equipment and land control results can lead to a decrease in the primary production and supply of plants, thus compromising the current and future potential. Micronutrition's mapping of soil is of utmost importance, as it is composed of traits such as pH, natural matter, and texture. These traits can be used to gain an understanding of the soil situation by observing what is happening at the bottom of the soil in terms of plant growth. Additionally, the water content material present within the soil is of great importance, and is taken into account through moisture mapping.

Flood Impact

Every year during the Kharif period, a significant portion of agricultural areas is affected by flash floods or excessive precipitation. Through the use of GIS, satellite imagery and ground-based statistics collected from ground survey teams, it is possible to calculate a specific harm assessment. In the event of a flood due to rainfall, the additional amount of water produced by the precipitation can have a detrimental effect on a rural area where no water drainage system is in place, resulting in inundation. Furthermore, a flood due to overflow can cause small tributaries to merge with the river, resulting in a flood scenario within the vicinity of riverbanks and river plains. This harm assessment of flood can significantly improve the effectiveness of land use planning in managing flood risk.

NATCAT Modeling

Natural Catastrophe Modeling is a tool used to calculate the actual time or potential outcome of chance evaluation, utilizing the probabilistic approach to predict the final outcomes and behavior of herbal hazards. It involves the mapping and measurement of risk through computer simulated disaster models, which incorporate clinical research and historical occurrences with the help of record generation and geographic records technology (remote-sensing and GIS). The flood chance maps are organized according to the predicted intensity of flood. This estimation is typically derived from hydrological and remote-sensed statistics, which ensure that regions with higher intensity of inundation can be assigned an excessive "risk denomination".

Drone Image Analysis for Crop Damage Assessment

Drone image evaluation has the potential to be highly beneficial in the evaluation of micro crops for hailstorm-induced crop loss, horticultural tree counting, disease-caused crop loss, and a variety of other factors. The spatial determinations of the input imagery are closely related to the precision of drone photograph statistics, as the spatial decision may begin at a distance of 50 cm and expand as necessary.

Crop Yield Prediction

Governments and agencies are increasingly relying on accurate yield prediction to ensure food security and to forecast earnings and budgeting. Recent advances in satellite communication, sensing, large-scale data collection, and artificial intelligence (AI) have enabled these predictions to be made more accurately. Convolutional neural networks (CNNs) are one of the most advanced strategies in this area. A ConvNet, or set of deep learning rules, is taught to identify the productivity of a crop by feeding it images of vegetation that has already been identified as productive. CNN's accuracy is estimated to be approximately 82%.

Crop Fitness Monitoring

Manually assessing crop fitness across multiple acres is the least efficient approach. This is where remote-sensing combined with GIS comes into play in farming. By utilizing satellite images and input data, environmental conditions can be evaluated across the area, such as humidity levels, air temperatures, surface conditions, and more. By utilizing GIS-based precision farming, this assessment can be improved and can help to determine which vegetation areas require the most attention.

Livestock Monitoring

Farm GIS software can be most effectively utilized in the area of animal husbandry by tracking the movements of specific animals. This allows farmers to locate animals on their farm and demonstrate their health, fertility, and nutritional status. The following GIS services can be utilized to achieve this goal: animal trackers, a mobile application to retrieve and display data from animal trackers, and virtual scales. Here is an example: You need to prove the weight of your beef cattle. Each animal has a tracker attached to its ear or neck, which verifies the animal's identification each time it is stepped on. This allows the machine to update its value for that identification.

Irrigation Control

Monitoring large fields to guarantee that each crop receives adequate water is a laborious task, but one that can be easily accomplished through the use of geoinformatization in agriculture. Helicopters and satellites equipped with high-resolution cameras capture images that enable AI algorithms to accurately calculate the water pressure in each crop and identify trends at the margins of water shortages.

Drought Assessment

The GIS application is used in drought assessment. The results of drought and NDVI show that vegetation condition can be used as indicator for drought condition of an area. NDVI can be used as the main indicator to evaluate drought. Difference images of NDVI show drought conditions In these images, if difference values are positive then area is less affected by drought, if difference values are zero the area is not affected by drought, and if difference values are negative then the area has affected by drought. Indices derived from the thermal bands, such as land surface temperature (LST) and the temperature condition index (TCI) have a higher sensitivity to drought conditions than indices derived from the visible bands, such as the NDVI, vegetation condition index (VCI) and soil adjusted vegetation index (SAVI). Low rainfall is usually the cause of drought, but high temperatures may also be involved. Drought as related to precipitation may be a result of several growing days without precipitation, low seasonal precipitation, or abnormally low annual precipitation In addition, the daytime and nighttime thermal data have been used to find the thermal inertia of the system and hence to interpret the moisture stress condition.

Challenges of Global Information Systems

Technological Challenges

- Not all nations have good enough data technology infrastructures
- Unable to construct worldwide information system
- Broadband communique traces needed
- Language is technological challenge
- Eight-bit bytes now no longer enough for languages with large character sets
- Unicode lets in for 65,536 characters
- Must coordinate with databases and applications

Regulations and Tariffs

- Countries have special importing regulations
- Executives reluctant due to hassles
- Comply with legal guidelines of destination nations
- NextLinx assist importers and exporters for web commerce

Language Differences

- International parties should agree on one common language
- Data now no longer transmittable across the world due to the fact data should be translated
- English taken into consideration as worldwide language
- Largest organizations translate websites into local languages

Political Challenges

- Some nations oppose policy of free access to statistics—gives other international locations possibility to manipulate indigenous resources.
- Government may require software program that to be purchased within borders
- Standards taken into consideration while integrating across the world
- Records can be incompatible
- United States makes use of English device of standard measures.
- Rest of world makes use of metric gadget
- *Different standards*:
    - Communicating dates
    - Times
    - Temperatures
    - Addresses
- United States makes use of month/day/year format
- Rest of the world makes use of day/month/year
- Opportunities provided with the aid of using GIS

Disadvantages

- Expensive as GIS could be very pricey software program to acquire
- A lot of technical knowledge is required for efficient operation of GIS as GIS machine has commonly complex command language that results in huge errors when it is dealt with the aid of a non-GIS employee
- It could be very tough to create consumer friendly as well as fast GIS packages

- For GIS outcomes to be accurate the inputted facts have to be correct, subsequently the GIS employees need to be aware of the truth that now no longer all the facts are usually accurate
- For correct outcomes of GIS accurate input number with exact data is prerequisite

Conclusion

Geographical information system is a cutting-edge technology that provides a wealth of research potential. It can be leveraged to assist decision makers in addressing challenges that were previously impenetrable due to the fact that the majority of agricultural issues are spatial in origin. For instance, there is a heightened demand for information on the management of GIS during the implementation and operational phases. Research is also required to examine the impact of GIS on organizations, collaboration, and decision-making efficiency as well as its impact on society in both developed and developing nations. As GIS becomes increasingly integrated into many information systems, IT researchers should be prepared to contribute their expertise to gain a more comprehensive understanding of the technology.

References

Abdelbaki, C., Benchaib, M.M., Benziada, S., Mahmoudi, H., Goosen, M. "Management of a water distribution network by coupling GIS and hydraulic modeling: a case study of Chetouane in Algeria." *Applied Water Science* 7, 3 (2017): 1561-7.

ESRI. 2016. An introduction to the commonly used GIS tools. https://desktop.arcgis.com/en/arcmap/latest/analyze/commonly-used-tools/an-introduction-to-the-fundamental-tools.htm [Last accessed February, 2024].

ESRI. 2022.. History of GIS—Timeline of Early History and the Future of GIS. https://www.esri.com/en-us/what-is-gis/history-of-gis [Last accessed February, 2024].

Pedamkar P. 2023. What is GIS? Top 4 Applications of Geographic Information Systems. https://www.educba.com/what-is-gis/ [Lats accessed February, 2024].

Şahin, Ş., Kurum, E. "Erosion risk analysis by GIS in environmental impact assessments: a case study—Seyhan Köprü Dam construction." *Journal of Environmental Management* 66, 3 (2002): 239-47.

Sarma, P.K. 2021. Components of GIS. Mukherjee University.

Grind GIS. 2017. Components of GIS. https://grindgis.com/blog/components-of-gis [Last accessed February, 2024].

Wiegand, J. "A guideline for the evaluation of the soil radon potential based on geogenic and anthropogenic parameters." *Environmental Geology* 40, 8 (2001): 949-63.

Yomralioglu, T., Colak, E., Aydinoglu, A. "Geo-Relationship between Cancer Cases and the Environment by GIS: A Case Study of Trabzon in Turkey." *International Journal of Environmental Research and Public Health* 6, 12 (2009): 3190-204.

2

Global Positioning System

Abstract

Smallholder and marginal farmers in developing countries like India need effective extension support systems to cope with the changing information needs and diversified production systems. Geotechnical technologies such as Global Positioning System (GPS) can help collect and analyze field data quickly and meet various information needs. Precision farming techniques can help increase crop yields, reduce production costs, and protect the environment. These techniques use GPS and geographical information system (GIS)-based applications to measure factors that affect crop yields, anomalies, and variations in management practices. The combination of GIS technology and relational databases can help visualize and present the impact of precision agriculture projects for small-scale and marginal farmers. This approach relies on the integration of multiple technologies such as GPS, miniaturized computing components, automated control, field, and remote-sensing, mobile computing, advanced information processing, telecommunications, etc. This decision support model can explain the PPP convergence of people (farmers), projects (farms), and pixels (digital images related to fields and crops). This information can help classify and group production systems for better service delivery. Precision agriculture is also known as GPS-based agriculture, variable-rate agriculture, prescription agriculture, site-specific agriculture, etc. Some forward-looking farmers have started using GPS to record their observations. In the future, GPS systems can help Indian farmers benefit from advanced technology without compromising their land or produce quality. This is the basis for developing a precision agriculture model for developing countries like India.

Keywords: Agriculture, satellite, GPS, geographic position, remote-sensing

Introduction

Global Positioning System (GPS) is a system of satellites and ground stations that provide location and time information for any place on earth, regardless of the weather. GPS is a type of GNSS, which stands for global navigation satellite systems. GPS has many applications in precision farming, such as farm planning, field mapping, soil sampling, tractor guidance, crop guidance, variable-rate applications (VRAs) and yield mapping. GPS helps farmers work in low-visibility conditions such as rain, dust, fog, and darkness. India has its own GPS system called IRNSS or NavIC, which uses about seven satellites. GPS is owned and operated by the U.S. government, specifically the U.S. Space Force. GPS satellites serve both civilian and military users around the world. The military service of GPS is available to U.S. and allied forces and authorized government agencies. GPS is a navigation system that uses >20 satellites and several ground facilities to provide accurate position, velocity, and time data in three dimensions, at any time and anywhere in the world. GPS is funded and controlled by the U.S. Department of Defense (DOD), which is responsible for designing and running the system for the U.S. military. GPS works by sending specially coded signals from satellites to receivers, which can calculate position, velocity, and time. To do this, a receiver needs signals from at least four GPS satellites.

Explanation

Global Positioning System is a system that uses signals from satellites to find a location on the earth's surface, no matter what the weather is. GPS relies on satellites that orbit the earth and send signals with their time and location. A receiver on the ground that gets signals from at least four GPS satellites can use math equations to figure out its location on the earth's surface. If the receiver keeps getting signals, it can also update its speed and direction if it is moving. GPS was first used for surveying and mapping because it gives a position in latitude and longitude without having to measure angles and distances between points. GPS is a complex system that can tell a person or a device where they are on the earth at any time and anywhere. GPS can find a position with different levels of accuracy, from a few millimeters to 100 m (El-naggar 2011).

Concept

Global Positioning System is a system that uses the time and position of satellites to find the location of a receiver on Earth. The satellites have very accurate atomic clocks that are synchronized with each other and with clocks on the ground. The satellites correct any errors in their time and position every day. The satellites send data about their time and position to the receiver constantly. The receiver uses multiple satellites and math equations to figure out its location and how much its clock differs from the satellite time. The receiver needs to see at least four satellites to calculate four things: its three coordinates and its clock error. There are always at least four satellites that the receiver can see from anywhere on Earth. The receiver gets radio waves from the satellites that tell it their time and position. The receiver can measure how long it takes for the radio waves to reach it from each satellite. This tells the receiver how far it is from each satellite. The receiver can then use a process called trilateration to find its location by using the distances from at least three satellites.

Trilateration

Trilateration is a way of finding a location by using the intersections of three or more spheres. The spheres are formed by measuring the distance from a satellite to a receiver, with the satellite at the center of the sphere. Trilateration can be used to calculate the time of sunrise, sunset, speed, and the distance from the receiver to the destination. Trilateration surveying is a method of measuring the sides of a triangle with electronic devices, and then computing the angles from them. A surveyor can use this method to find other distances and angles that are hard to measure by making a series of triangles next to each other (Carter 1997). Trilateration used to be less common than triangulation, which is a method of measuring angles instead of sides, but it became more popular with the development of electronic distance-measuring devices.

Triangulation

Triangulation is the main principle behind the GPS's amazing navigation ability. Triangulation works by measuring the time it takes for a satellite signal to reach the earth, which is less than a tenth of a second. Then, it multiplies that time by the speed of a radio wave, which is 300,000 km (186,000 miles) per second. This gives the distance between the satellite and the receiver. This distance puts the receiver on an imaginary sphere with the satellite at the center. By doing the same thing with three more satellites (four in total), the receiver's computer can find the point where all four spheres meet. This point is the user's current location in longitude, latitude, and altitude. The receiver can also calculate the user's current speed and direction by measuring how the signals from the same four satellites change in frequency due to the Doppler effect caused by their combined motion (Hoque 2016).

Augmentation

The satellite signal travels very fast to the Earth, but a lot can happen to it in that short time. Farmers who use these signals can guide their field equipment very accurately, which is why precision farming is a common term in agriculture. The GPS augmentation technique uses carrier waves that carry the satellites' navigation pulses to the Earth. The carrier wave is much shorter and more frequent than the basic navigation pulse, so this technique can reduce navigation errors to <1 cm (0.4 inch) under the right conditions (Sarkar 1997). This technique takes advantage of the shorter and more numerous carrier waves that hit the receiver's antenna every second. Another augmentation technique is geosynchronous overlay, which uses GPS payloads that are attached to commercial communication satellites in geostationary orbit, about 35,000 km (22,000 miles) above the Earth. These small payloads send the civilian C/A-code pulse trains to the users on the ground.

Applications

Global Positioning System is a system that uses satellites to find your location on earth. It can do many things for agriculture, such as:

- *Giving a location*: It can tell you where you are in coordinates, like latitude and longitude or Universal Transverse Mercators (UTMs)
- *Point to point navigation*: It can help you create a route from one point to another by using waypoints. You can use a map, the coordinates of a place, or your current location to make waypoints
- *Route navigation*: It can help you move along a trail by using multiple waypoints. It can show you the direction and distance to the next waypoint
- *Surveying and mapping*: It can help you measure the coasts and waterways, where there are few landmarks. Survey vessels use GPS and sonar to make the charts that show the water depths and underwater dangers
- *Soil sampling*: It can help you find out the soil quality and suitability for different crops. Soil sampling can also help you make profiles of soils and separate the good ones from the bad ones
- *Weed detection*: It can help you locate the patches of weed by using linear sampling. Weeds can reduce the crop growth and yield over time
- *Proper planting*: It can help you plant crops with the right space and depth for each seed. Different seeds have different spacing and depth requirements depending on the soil type. GPS can advise you on the best planting practices for maximum yield
- *Planting ratios*: It can help you decide how many seeds to plant per area. Some crops need more space than others, while some crops can be planted together. GPS can help you determine the ratio of this type of planting

- *Yield maps*: It can help you make maps of the expected yield for different crops. For example, during harvests, GPS can map out how much crop you can get per area based on the land and seed characteristics
- *Harvesting calculations*: It can help you decide when and how to harvest your crops. It can also estimate the size of the area being harvested and the expected returns per area
- *Yield map location*: It can help you find the yield map by putting a GPS receiver on the farm machinery and collecting data
- *Environmental control*: It can help you apply pesticides, herbicides, or other chemicals based on the needs of each square meter. This reduces the amount of chemicals used and prevents waste and runoff
- *Farm planning*: It can help you plan your farm before planting. It can tell you the size of the area and help you decide what crop to plant on what part of the farm by using factors like soil and crop characteristics
- *Field mapping*: It can help you estimate the field that you want to farm. This way, you can decide which part of the field is good for farming and which part is not
- *Soil sampling*: It can help you know the soil type and quality on your farmland. This can help you choose the best crop for your soil
- *Crop scouting*: It can help you map an area and find the best crops for it. Scientists can use this to improve the crop quality by producing crops that suit the location
- *Correlation studies*: It can help you compare the crop yields from different production techniques. You can use this information to evaluate the effectiveness of each technique
- *Soil property mapping*: It can help you determine the soil properties and variations for different crops. You can use this to identify the soil types and suitability in different parts of your farm
- *Machinery spotting*: It can help you locate your farm machinery on a large piece of land. You do not have to go out and look for your equipment manually, especially if you have many of them
- *Machinery direction*: It can help you use autonomous farm machines for planting. It can direct these machines to decide where and how to plant the seeds with the right spacing and depth

- *Classification of areas*: It can help you classify the areas for cultivation based on factors like soil types and terrain maps. You can separate the areas that are good for farming from those that are not
- *Assessment of water availability*: It can help you find the water sources in your area, such as rivers or canals. You can use GPS to locate them easily
- *Identification of irrigated crops*: It can help you distinguish between irrigated and nonirrigated areas. You can use this to compare the irrigated crops and nonirrigated crops
- *Identification of swamps/waterlogged areas*: It can help you identify swampy and waterlogged areas that may not be good for some crops. You can use this to decide which crops are suitable or unsuitable for different lands
- *Rivers mapping*: It can help you map all the rivers in your area, which shows the water flow. Farmers and scientists can use this to know the presence of rivers and choose the crops that can grow well in the river basins
- *Land usage pattern*: It can help you monitor how your land is used. You can use GPS to know how much of your land is cultivated and how much is left bare
- *Contour mapping*: It can help you make contours in hilly or uneven terrains. You can use GPS to do this in a specific area
- *Irrigation systems mapping*: It can help you locate some irrigation systems like dams and canals. This can make it easier to find out how to irrigate your crops from the water sources
- *Meteorological mapping*: It can help you map some weather conditions that may affect the type of crop that can grow in a region
- *Personnel mapping*: It can help you map how many workers are on your farm at a certain time of the day. This is useful when you want to measure the productivity of your workers on a farm
- *Plantation mapping*: It can help you map a plantation crop and estimate its yield
- *Water bodies mapping*: It can help you map the water bodies in your area and assess their impact on crop growth and yield
- *Geofencing*: A geofence is an invisible fence that triggers a response when a GPS device crosses it. Geofences are used to protect your

vehicles and equipment from theft, or to alert you of restricted or sensitive areas. Geofences also help you mark areas that need attention, such as fertilizing or harvesting, or areas that have been treated with chemicals or are organic

- *Track farm equipment and workers*: GPS tracking devices and fleet management solutions help you keep track of where your vehicles and equipment are located. GPS tracking also helps you know where your field workers are at all times, which improves communication and decision-making. GPS tracking also reduces supervision, which means fewer workers can manage larger areas, saving labor costs
- *Theft protection*: GPS tracking devices and geofences can prevent or recover stolen farm vehicles and equipment. You will always know where your assets are, and if one is stolen, you can easily track it down with GPS tracking
- *Improve farming routes*: GPS tracking devices and fleet management solutions help you optimize your farming routes. You can avoid wasting time on land that has already been tended to, or find the best route to reach your destination. This also saves fuel and reduces wear and tear on your vehicles and equipment
- *Optimize farm equipment life cycles*: GPS tracking devices and fleet management solutions help you prolong the life of your equipment by managing maintenance schedules and monitoring diagnostic information. You can prevent costly repairs by fixing small problems early, or negotiate higher resale prices by showing that your assets were well-maintained

History

- The 1970s saw the implementation of phase I—the concept validation phase, during which the first prototype satellites were manufactured and tested
- *In 1973*: Development of a satellite navigation system for military
- *In 1974–1979*: System tests for US air force and navy; the first functional NAVSTAR prototype satellite launch occurred in June 1977, and was called the NTS-2 (Navigation Technology Satellite 2, which was actually a modified Timation satellite) While the NTS-2 only survived some 7 months, the concept was shown to be viable, and in February, 1978, the first of the Block I NAVSTAR satellites was launched

- In 1977, first receiver test was conducted without placing the satellite in the orbit. The signal was received from pseudosatellites. The concept started in the late '60s but the first satellite was launched during February 1978
- In 1979, phase II—full-scale development and testing of the system, was implemented with nine more Block I satellites launched during the following 6 years
- This was followed in late 1985 by phase III—the full-scale production and deployment of the next generation of Block II satellites
- In 1978–85, total 11 Block I satellites were launched
- *In 1979*: decision was taken to expand GPS with 18 satellites in space
- In 1980–1982, financial crisis felt when the sponsors put questions on the usefulness of the system
- In 1983, the civilian use of GPS was permitted after Soviet Union shot down Korean airplane that was lost over soviet territory
- In 1986, GPS program suffered a setback due to the accident of challenger, which was supposed to carry Block II satellites to the orbit. Then delta rockets were used for the purpose
- In 1988, numbers of satellites get increased to 24
- In 1989, first Block II satellites installed and activated also; the Magellan Corp. introduced the first handheld GPS receiver
- In 1990–1991, during the Gulf War temporal deactivation of selective availability (SA)
- In 1992, GPS was put to use in "Operation Desert Storm 1993—Initial operation capability (IOC) was announced and decision was taken worldwide that civilian use free of cost
- In 1994, the last Block II satellites complete the satellite constellation
- In 1995, fully operational capability was declared
- On March, 1996, the president made decision for civilian use of GPS, i.e., free
- In 2000, final deactivation of SA, providing positional accuracy of 20 m from 100 m
- In 2005, the launch of the II RM GPS satellite which supports the new military M signal and the second civil signal L2C

- In 2013, Indian Regional Navigation Satellite System was launched by ISRO (Indian Space Research Organisation) (Mr Ershad Ali, Department of Geography, Ananda Chandra College, Jalpaiguri; April, 2020)

Approach

Global Positioning System is a system that uses satellites to find your location on earth. It was originally developed for military purposes, but it has also been used for many nonmilitary purposes over the years. One of the industries that has benefited from GPS is agriculture. GPS can help farmers improve their efficiency and productivity in many ways. By 2050, the world's population is expected to reach nearly 10 billion people. This means that farmers need to produce more food to meet the demand. GPS can help farmers use precision agriculture and advanced technologies such as automation, computer vision, artificial intelligence (AI), and machine learning to achieve this goal. Precision agriculture is a system that uses GPS, drones, and satellite images to collect data on various factors, such as crop status, weather forecasts, environmental changes, and so on. Precision agriculture also allows farmers to manage their fields as separate zones rather than as a single unit. This way, farmers can make more specific decisions for each part of the field, such as adjusting fertilizer amounts, optimization techniques, and fuel consumption (Shanwad et al., 2002).

Components

Space Segment

The GPS system has three main parts: satellites, ground stations, and receivers. Satellites are the space segment of the system. They orbit the earth at a height of 12,000 miles and send signals that tell their position and time. There are 24 satellites in six different orbits, each tilted at 55° to the equator. This way, a receiver on earth can always see at least four satellites in the sky. Satellites complete one orbit every 12 hours and pass over the same spot on earth every 24 hours (Samama 2008). The ground stations are the control segment of the system. They monitor and control the satellites, and send them updates on their orbits and clocks. The ground stations also collect data from the satellites and use it to predict their future behavior. The predictions are then sent back to the satellites for them to broadcast to the users. The ground stations make sure that the satellites stay on track and do not drift too far from their planned orbits (Massatt et al., 2003).

Control Segment

The GPS system has a control segment that consists of stations around the world that track and manage the satellites. The stations make sure that the satellites stay in their correct orbits and have accurate clocks. They also send updates to the satellites with new navigation data and status information. The control segment has one main station, one backup station, 12 antennas, and 16 monitors. The satellites are the space segment of the system. They circle the earth at a height of 17,440 km and take 12 hours to complete one orbit. They are angled at 55° to the equator, so that they can cover the whole earth, including the poles. The satellites have solar panels that face the sun and antennas that face the earth. They also have four atomic clocks that keep very precise time. The satellites send signals that contain their position, time, and other data. The receivers use these signals to calculate their own location and time. The receivers measure how long it takes for the signals to reach them from the satellites. By knowing the distance to at least four satellites, a receiver can find out its latitude, longitude, and altitude.

User Segment

The user segment of the GPS system is the receiver equipment that picks up the signals from the satellites and uses them to figure out the user's location and time. The receiver equipment can be different in price and design, depending on how it is used. Most receivers have three main parts:

- *Antenna*: Gets the signal and can block interference
- *Receiver–processor unit*: Changes the radio signal to navigation data
- *Control/display unit*: Shows the location data and lets the user control the receiver

Types

- *A-GPS*: Assisted GPS or A-GPS is a kind of GPS that uses local network sources to help receivers find satellites. It is useful in places where satellite signals are blocked by trees or buildings. But it needs cellular networks to work. A-GPS was developed when GPS was used for commercial purposes. A-GPS is common in smartphones, where it lets emergency services locate the caller. It also makes the phone connect to the navigation system faster and better, even with weak signals or few satellites
- *S-GPS*: Simultaneous GPS or S-GPS is a modified GPS that allows phones to send voice data and GPS signals at the same time. This means that both types of data are transmitted together, not one after another This improves the sensitivity and quality of the signals. This is helpful in emergencies, because it lets service providers like ambulances or fire stations find the caller while talking to them

- *D-GPS*: Differential GPS or D-GPS is a correction technique that improves the accuracy of location data from a normal GPS receiver. D-GPS is an upgrade to GPS, giving a more precise idea of where something or someone is. The usual accuracy for GPS is about 15 m at best. D-GPS can reduce the error limit to 1 inch (2.5 cm). It works by using a network of fixed ground stations that send the difference between their known locations and the satellite locations. D-GPS is used by the United States and the Canadian Coast Guards
- *Nondifferential GPS*: Nondifferential GPS, unlike differential GPS, uses direct satellite signals to figure out the position. It is less accurate than D-GPS, but it can be used in more places
- *Mapping and nonmapping GPS*: Mapping GPS is a type of GPS unit that has built-in maps. It can also download more maps to add to the unit. This is the type of GPS unit that is often found in mobile devices and other handheld devices. Nonmapping GPS is a type of GPS unit that does not have maps. It shows the location and the direction to another point without showing roads or landmarks. It uses trails like breadcrumbs to mark the way

Methodology

Variable-rate application technology spreads precise amount of product based on location where controller linked with a field map by using data of plant biomass, soil nutrient status, or plant green leaf area with the help of position locator input can be used precisely. There are two methods of VRA technology, depending on whether it uses GPS system or not:

- *Map-based VRA*: This method uses a GPS receiver to adjust the application rate according to an electronic map, also called a prescription map. A prescription map is an electronic data file that contains specific information about the input rates for each zone in the field
- *Sensor-based VRA*: This method does not need GPS or prescription maps. The sensors on the applicators measure the soil properties and crop characteristics

Tools and Techniques

Precision agriculture or site-specific farming is made possible by the combination of the GPS and GIS. These technologies allow the collection of real-time data with accurate location and information, leading to the efficient analysis and management of large amounts of geospatial data. GPS-based applications in precision farming are used for various purposes, such as farm planning, field mapping, soil sampling, tractor guidance, crop scouting, VRAs, and yield mapping. GPS also enables farmers to work in low-visibility conditions, such as rain, dust, fog, and darkness. In the past, farmers had difficulty in correlating production techniques with crop yields due to land variability. This limited their ability to develop the best soil/plant treatment strategies that could have improved their production. Today, precision agriculture enables more precise application of pesticides, herbicides, and fertilizers, and better control of their dispersion, reducing costs, increasing yield, and creating a more environmentally friendly farm. Precision agriculture is changing the way farmers and agribusinesses view the land from which they earn their profits. Precision agriculture is about collecting timely geospatial information on the soil–plant–animal needs and then prescribing and applying site-specific treatments to increase agricultural production and protect the environment. Instead of treating their fields uniformly, farmers are now seeing benefits from managing their fields at a finer scale. Precision agriculture is gaining popularity largely due to the introduction of high technology tools for the agricultural community that are more accurate, cost-effective, and user friendly. Many of these new innovations rely on the integration of on-board computers, data collection sensors, and GPS time and position reference systems. Many people think that the benefits of precision agriculture can only be realized on large farms with huge capital investments and experience in information technologies. But this is not true. There are some inexpensive and easy-to-use methods and techniques that can be developed for farmers' use. By using GPS, GIS, and remote-sensing, information required for improving land and water use can be obtained. Farmers can also achieve additional benefits by combining better utilization of fertilizers and other soil amendments, determining the economic threshold for treating pest and weed infestations, and preserving the natural resources for future use.

Challenges

- *Increasing GNSS signal coverage*: Many farm environments have poor GPS coverage due to hills, mountains, or trees that block the signal. Most high-precision systems in North America now use GLONASS to supplement GPS for better signal coverage. As more satellites are launched, GNSS will become more dependable in difficult environments and more appealing to growers
- Current methods of measuring soil nutrients are slow, costly, and imprecise. Many growers use zone sampling techniques, which take only one to five samples per field (usually 80–160 acres) based on soil

texture zones. Some growers use high-density sampling techniques, which take one sample per 2.5 acres

- Crop yield is influenced by many factors, such as rainfall, temperature, humidity, wind, soil type, and tillage practices. A single farm cannot produce enough data to make meaningful decisions. Data analysis will improve significantly when data can be compared across many farms. Wireless data connections and software tools that allow such analysis are just starting to appear on farms. Once they become more common, the value of precision farming will become more evident to growers and adoption will increase
- Technology plays an increasingly important role in agriculture, and original equipment manufacturers (OEMs) have been riding a wave of innovation that has brought many benefits to the industry. However, being at the forefront of innovation also comes with some challenges
- Connectivity is a key challenge for the future of agriculture, as farmers increasingly use the Internet of Things to enhance their productivity. Many new technologies in precision agriculture require a high-speed internet connection—at least 4G. Without good connectivity, farmers will face difficulties in adopting new technologies and advancing precision agriculture. The growth of digital farming depends on the pace of connectivity improvement
- Agriculture can vary in size, from individual farmers to large corporations that manage hundreds of hectares. They often face similar problems and need similar solutions. For precision agriculture to become widespread, the technology that farmers use must be adaptable to their own scale. Whether they have personal gardens or industrial farms, they should be able to use the same tools for precision farming

Advantages

Precision agriculture has many benefits for farmers and the environment. Some of these benefits are as follows:

- It allows farmers to map the yield and soil characteristics of their fields and divide them into smaller plots based on their specific needs
- It helps farmers to manage their resources better and reduce waste by applying the optimal amount of agrochemicals and water to each plot
- It minimizes the environmental risk of nitrate leaching and groundwater contamination by using agrochemicals more efficiently

- It enables farmers to collect and analyze data on soil, crop, and weather conditions and adjust their planting density and chemical applications accordingly
- It improves field navigation and reduces overlapping and skipped areas by using GPS and other technologies. This also saves time and increases productivity
- It allows farmers to work in low visibility conditions such as rain, fog, dust, and darkness without compromising accuracy and safety
- It eliminates the need for human operators in spraying operations and reduces overspray and drift
- It monitors the yield data and helps farmers to plan for future site-specific field preparation
- It spreads information about agricultural practices that can improve the quality, quantity, and cost-effectiveness of crop production
- It enhances agricultural productivity and prevents soil degradation, leading to sustainable agricultural development

Disadvantages

- The technology is still evolving and requires expert guidance before making costly decisions
- The initial investment may be high and it may take a long time to see the returns
- It may require several years of data collection and analysis to fully implement the system
- The work is very demanding, especially in terms of data management and interpretation

Conclusion

Global Positioning Systems help farmers to use inputs (seed and fertilizer) more wisely, to operate tillage equipment more smoothly, to measure crop and field conditions more precisely, and to make better farm management decisions by using electronics, computers, software, and sensors, farmers now have better tools to manage their crop production. Some of these tools, such as seed genetics and precision steering, have been rapidly adopted by farmers in the United States, because they are easy to see and use. However, farmers are still reluctant to adopt seed and fertilizer management tools. This is a serious problem because using the wrong amount of fertilizer can harm both the farmers' income and the environment. Farmers will adopt GPS systems for seed and fertilizer management more eagerly when three key challenges are overcome: improving the reliability of GNSS signals, improving the speed of soil measurements, and enabling data analysis across multiple farms.

References

Carter, C. 1997. "Principles of GPS Positioning." *A Brief Primer on the Operation of the Global Positioning System. CSR Revision*, volume. 4900. United States: Allen Osborne Associates.

El-naggar, A.M. "An alternative methodology for the mathematical treatment of GPS positioning. *Alexandria Engineering Journal* 50, 4 (2011): 359-366.

Hoque, M.Z. "Basic Concept of GPS and Its Applications." *IOSR Journal Of Humanities and Social Science (IOSR-JHSS)* 21, 3 (2016): 31-7.

Massatt, P., Fritzen, F., Perz, M. *Assessment of the Proposed GPS 27-Satellite Constellation.* Portland: Proceedings of the 16th International Technical Meeting of the Satellite Division of The Institute of Navigation (ION GPS/GNSS 2003). pp. 399-406.

Samama, N. 2008. *Global Positioning: Technologies and Performance*. Germany: Wiley. p. 419.

Sarkar, A. 1997. *Practical Geography: A Systematic Approach*. New Delhi: Orient BlackSwan.

Shanwad, U.K., Patil, V.C., Dasog, G.S., Mansur, C.P., Shashidhar, K.C. 2002. *Global Positioning System (GPS) in Precision Agriculture*. Karnataka: Proceedings of Asian GPS Conference.

3

Remote-Sensing

Abstract

Remote-sensing is a technique that uses electromagnetic radiation to measure the characteristics of objects without touching them. A satellite remote-sensing system has five parts, namely, radiation sources (such as the sun, the earth, or an artificial source), atmospheric interaction, surface interaction, space segment (sensors), and ground segment. The sensors in remote-sensing do not contact the objects or events they observe directly. They need a physical carrier to transmit the information from the objects/events to the sensors through a medium. The carrier is electromagnetic radiation in remote-sensing. Remote-sensing instruments in space allow us to observe the earth's surface changes, natural resources management, and environmental conditions. It has many applications in crop identification, yield monitoring, crop damage assessment and can work at any time of the day.

Keywords: Electromagnetic spectrum, first commercial satellite (IKONOS), GIS analysis, Landsat, LIDAR, radar, radiation, wavelengths

Introduction

We use our five senses to perceive the world around us. Some senses, like touch and taste, need contact with the objects. But we get most of our information about the environment through sight and hearing, which do not need close contact with the external objects. In scientific terms, we are doing remote-sensing all the time. Remote-sensing is the process of recording/observing/perceiving (sensing) objects or events from a distance. Remote-sensing is a way of monitoring and detecting the features, usually physical, of a specific area by measuring its reflected and emitted radiation (usually from satellite or aircraft). There are special cameras that capture remote sensed images, which help researchers "sense" things about the earth. Remote sensors collect data by detecting the energy that is reflected from earth. These sensors can be on satellites or mounted on aircraft (Joseph and C 2018). The discipline of remote-sensing has emerged as one of the most mesmerizing subjects

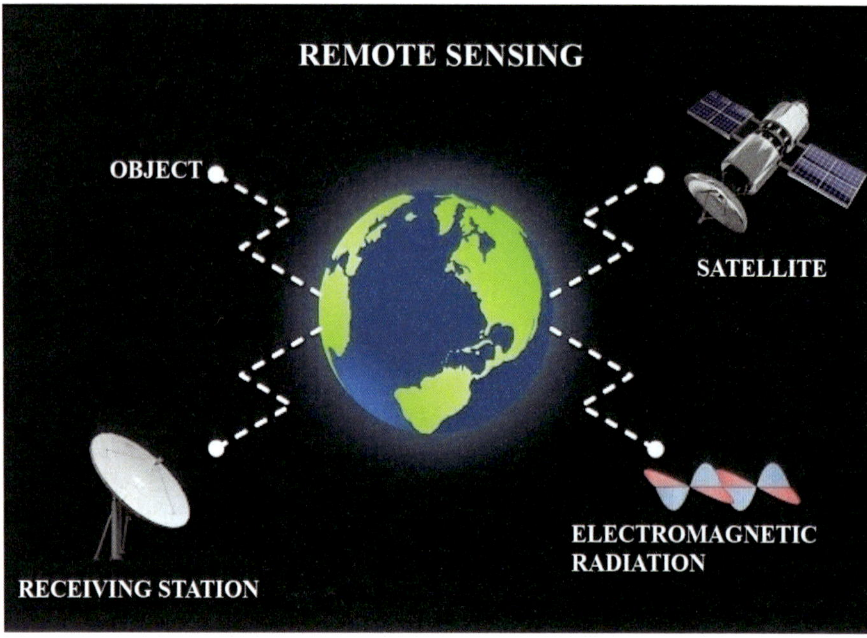

Fig. 1: Overview of remote sensing.

over the past three decades. Earth observation from space through various remote-sensing instruments has provided a vantage means of monitoring land surface dynamics, natural resources management, and the overall state of the environment itself (Gitelson et al., 2002) (Fig. 1).

Principle

Remote-sensing works on the basis of how different objects interact with the electromagnetic spectrum and how they are positioned relative to the satellite or airborne platforms that observe them. To identify or detect objects or surface features, amount of energy they reflect or emit in different parts of the electromagnetic spectrum is to be measure. This depends on the material properties (structural, chemical, and physical), surface roughness, angle of incidence, intensity, and wavelength of the energy. Each object has a different reflectance or emittance pattern at different wavelengths, which is unique to that object and is called its spectral signature (Gitelson et al., 2002). A human interpreter does not use the spectral information as well as a computer. Humans can see color differences very well, but they can only tell apart a few levels of brightness (Cihlar and Protz 1972).

Remote-Sensing System/Methodology

To understand remote-sensing better, we can break it down into different steps. They are:

1. The source of electromagnetic energy (sun, a transmitter carried by the sensor)
2. The travel of energy from the source to the earth's surface and how it interacts with the atmosphere in between
3. The interaction of energy with the earth's surface (reflection/absorption/transmission) or self-emission
4. The travel of the reflected/emitted energy to the remote sensor on a suitable platform, through the atmospherc in between
5. The measurement of the energy by the sensor, converting it into a photographic image or electrical output
6. The transmission/recording of the sensor output
7. The preprocessing of the data and generation of the data products
8. The collection of ground truth and other related information
9. The analysis and interpretation of the data
10. The integration of interpreted images with other data to derive management strategies for various themes or other applications

Components

Remote-sensing often involves the interaction between incident radiation and the targets of interest. This process can be described by the following seven elements:

1. *Energy source or illumination*: The first element of remote-sensing is to have an energy source that provides or emits electromagnetic energy to the target of interest
2. *Radiation and the atmosphere*: As the energy travels from its source to the target, it interacts with the atmosphere it passes through. This interaction can also happen again as the energy travels from the target to the sensor
3. *Interaction with the target*: When the energy reaches the target through the atmosphere, it interacts with the target depending on the properties of both the target and the radiation

4. *Recording of energy by the sensor*: After the energy is scattered by or emitted from the target, we need a sensor (remote, not in contact with the target) to collect and record the electromagnetic radiation
5. *Transmission, reception, and processing*: The energy recorded by the sensor has to be transmitted, usually in electronic form, to a receiving and processing station where the data are processed into an image (hardcopy and/or digital)
6. *Interpretation and analysis*: The processed image is interpreted, visually and/or digitally or electronically, to extract information about the target that was illuminated
7. *Application*: The last element of remote-sensing is achieved when we use the information we have extracted from the imagery about the target to understand it better, reveal some new information or solve a problem

Types of Remote-Sensing

- *On the basis of energy*:
  - Active: RADAR is a type of active remote-sensing that uses energy to scan targets and measure their properties. It sends out radiation and then receives the radiation that bounces back from the objects or areas. By calculating the time difference between the emission and the return, it can determine the location, height, speed, and direction of a target
  - Passive: Passive remote-sensing observes objects or areas without sending out any radiation. It relies on the sunlight that is reflected by the targets. Film photography, infrared, and radiometer are some examples of passive remote-sensing devices
- *On the basis of orbit*:
  - Polar: It is also known as sun synchronus. It goes around the poles at an altitude of around 1,000 km and orbital period of 100 minutes. It is used for earth observation, earth mapping and reconnaissance survey, and monitoring weather
  - Geostationary: It is stationed at an altitude of 360,000 km above the earth looking at same place always
- *On the basis of range of electromagnetic spectrum*:
  - Optical remote-sensing: Remote-sensing is visible near infrared and middle infrared ranges (0.3–3 mm)

- Thermal remote-sensing: It emitted from 3 to 5 mm and 8 to 16 mm is sensed
- Microwave remote-sensing: Remote-sensing in higher wavelengths from 1 mm to 1 cm. This larger wavelength can penetrate even clouds

Electromagetic Spectrum

The electromagnetic spectrum has different regions of radiation with different wavelengths. The shortest ones are gamma and X-rays, and the longest ones are microwave and radio waves. Some regions of the electromagnetic spectrum are useful for remote-sensing, which is the observation of objects or areas from a distance. One of these regions is the visible spectrum, which is the light that we can see with our eyes or with remote sensors. However, the visible spectrum is very small compared to the rest of the electromagnetic spectrum. There are many other types of radiation that we cannot see, but we can use remote-sensing instruments to detect them and use them for various purposes. The visible spectrum ranges from about 0.4 to 0.7 micron in wavelength (Table 1). The longest wavelength in the visible spectrum is red, and the shortest one is violet.

Software Used in Remote-Sensing

As shown in Figure 2, remote-sensing data are processed and analyzed with computer software, known as a remote-sensing application.

Applications of Remote-Sensing

- *Analyzing rural road conditions*: Remote-sensing and geographic information system (GIS) techniques can be used to assess the quality of rural roads with high accuracy. This can save time and money for transporters

Table 1: Spectrum range and optical indicator.

Spectrum range (μm)	Optical indicator
0.4–0.446	Violet
0.446–0.500	Blue
0.500–0.578	Green
0.578–0.592	Yellow
0.592–0.620	Orange
0.620–0.777	Red

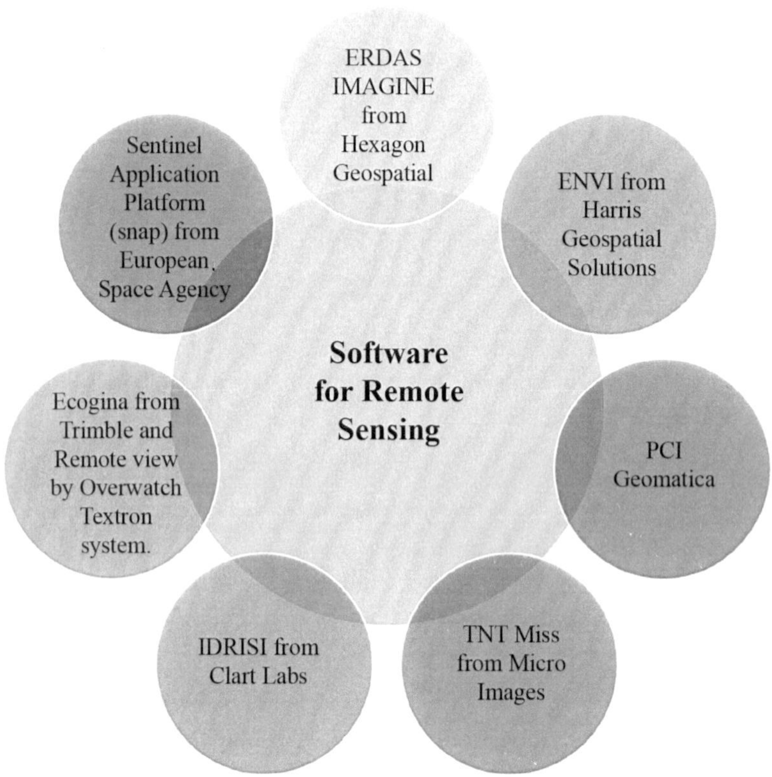

Fig. 2: Software for remote-sensing.
Source: Schowengerdt 2006

- *Creating base maps for visual reference*: Many modern mapping technologies rely on Remote-Sensing data, such as Google maps, open street maps, Bing maps, NASA's Globe view, and more
- *Measuring snow melt ratio*: Remote-Sensing technology can help understand how much snow melts over time. NASA uses LIDAR and a spectrometer to measure the absorption of sunlight by snow
- *Collecting earth's images from space*: Many space organizations have collections of images of earth. These images show interesting patterns of earth's geometry, atmosphere, oceans, land, and more. EO-1, Terra, and Landsat are some of the satellites that collect this data
- *Controlling forest fires*: Remote-sensing data from satellites can help firefighters to respond quickly and accurately to forest fires. This can reduce the damage caused by such fires

- *Detecting land use and land cover*: Remote-sensing technologies can be used to determine the physical characteristics of land and how it is being used (land use)
- *Estimating forest resources*: MODIS, AVHRR, and SPOT are some of the satellites that monitor the changes in global forests, which are vital sources of valuable materials such as paper, packaging and construction materials, and more
- *Locating construction and building alteration*: Tax revenue agencies use satellite data in some countries such as Greece and Athens. They use this data to find signs of wealth, like unclaimed swimming pools, and fight against tax evasion
- *Identifying fraud insurance claims*: Some insurance companies use Landsat's red and infrared channels to measure the vegetation growth in certain lands. This information can be used to verify the crops that are insured and prevent crop insurance fraud
- *Observing climate changes*: Satellites such as CERES, MODIS, AMSRE, TRMM, and MOPITT can observe the changes in climate from space. They can also compare the past and present climate conditions
- *Predicting potential landslides*: Landslides cause significant death and property loss around the world. INSAR uses interferometry remote-sensing technique to provide an early warning for possible landslides

Applications of Remote-Sensing in Agriculture

- *Yield monitoring*: Remote-sensing can help estimate the expected crop production and yield in a given area. It can also show how the crop will perform under different conditions. This can help researchers to forecast the amount of crop that will be harvested in a given season
- *Crop damage assessment*: Remote-sensing can help detect and measure the extent of crop damage or crop progress in a farmland. It can penetrate the crop canopy and show how much of the crop has been affected by pests, diseases, drought, or other factors
- *Crop identification*: Remote-sensing can help identify the type and characteristics of crops in a farmland. It can collect data from the crop and analyze it in the laboratories. It can reveal the crop culture and other features of the crop
- *Crop acreage estimation*: Remote-sensing can help measure the size of the land where a crop has been planted. This can be difficult to do

manually because of the large areas involved. Remote-sensing can provide accurate and reliable data on the crop area

- *Crop condition assessment and stress detection*: Remote-sensing can help evaluate the health and quality of each crop and how well it has coped with stress. It can show the crop vigor, nutrient status, water status, and other indicators of crop condition
- *Soil mapping*: Remote-sensing is widely used for soil mapping. It can help identify and classify different types of soils and their properties. It can also detect problematic soils that have low fertility or other issues that affect crop yield
- *Land cover and land degradation mapping*: Remote-sensing can help map out the land cover of a given area. It can show what types of vegetation, water, or other features are present on the land. It can also show the changes in land cover over time and identify areas that have been degraded or eroded
- *Pest and disease infestation identification*: Remote-sensing can help identify and locate pests and diseases that infest crops in a farmland. It can provide data on the type, severity, and distribution of the pest or disease. It can also suggest the best pest control methods to use to protect the crops (Hatfield et al., 2008)

Challenges in Remote-Sensing

- They are indirect measurements of the phenomenon, so they need to be compared with reality. This comparison is never perfect; a 10% error in classification is very good. They need to be adjusted geometrically and georeferenced to be useful as maps, and not just as pictures
- Different phenomena can be mistaken if they have the same appearance to the sensor, leading to classification error, for example, natural and artificial grass in green light
- Phenomena that were not intended to be measured can affect the image and must be taken into account
- Resolution of satellite imagery is too low for detailed mapping and for distinguishing small contrasting areas

Conclusion

Remote-sensing is the collection of information about the earth's surface without touching it or the object being studied. The methods include taking pictures from the air, using different types of light, and using radar. Remote-sensing can help us get accurate information about the earth's surface and its parts, such as forests, landscapes, water resources, oceans, and more. This information can help researchers to study the earth's parts and how to manage and protect them. To do remote-sensing, a sensor needs to be on a stable platform that is away from the target or surface being observed. The platform can be on the ground, on a plane or balloon (or something else in the earth's atmosphere), or on a spacecraft or satellite outside the earth's atmosphere.

References

Cihlar, J., Protz, R. "Perception of tone differences from film transparencies." *Photogrammetria* 28, 4 (1972): 131-40.

Gitelson, A.A., Kaufman, Y.J., Stark, R., Rundquist, D. "Novel algorithms for remote estimation of vegetation fraction." *Remote Sensing of Environment* 80, 1 (2002): 76-87.

Hatfield, J.L., Gitelson, A.A., Schepers, J.S., Walthall, C.L. "Application of Spectral Remote Sensing for Agronomic Decisions. *Agronomy Journal* 100, S3 (2008): S117-31.

Joseph, G., C., J. 2018. *Fundamentals of Remote Sensing*. India: Universities Press.

Schowengerdt, R.A. 2006. Remote Sensing. *Remote Sensing: Models and Methods for Image Processing*. Netherlands: Elsevier.

4

Generation of Spectral Profiles of Different Objects

Abstract

When energy hits the earth's surface, it can be reflected, absorbed, or transmitted depending on the wavelength and the features of the surface. For example, the surface could be soil, water, or vegetation. Reflection happens when energy bounces back from the surface at the same angle as its source. Absorption happens when energy is taken in by the surface. Transmission happens when energy passes through the surface. These three processes are not separate. The energy that hits a surface may be partly reflected, absorbed, or transmitted depending on the wavelength, features, angle, and properties of the surface. The energy that is reflected or reemitted from the surface is recorded by the sensors and analyzed to identify and interpret the features of the surface. If all the energy is absorbed by the surface and none is reflected, then the absorbance is one and the reflectance is zero. If all the energy is reflected by the surface and none is absorbed, then the absorbance is zero and the reflectance is one. The reflectance of a surface at different wavelengths is called spectral reflectance or albedo. It ranges from 0% to 100%. Different features of the earth's surface have different spectral responses or reflectances in remote sensing systems. Each feature or object has a unique spectral signature that can be used to identify it and study its properties.

Keywords: Spectral signature, Spectral reflectance, Spectral profiles, Reflectance

Introduction

Spectral signature: A spectral signature is the unique pattern of reflection and emission of a target at different wavelengths. This difference helps to classify remote sensing images based on the spectral signatures of different objects.

Electromagnetic (EM) spectrum: The EM spectrum is the range of all types of EM radiation. Some examples are visible light from a lamp, radio waves from a radio station, microwaves, infrared light, ultraviolet light, X-rays, and gamma rays.

Spectral response patterns: A spectral response pattern is the amount of energy that an object reflects or emits at different wavelengths. It shows how the object interacts with the EM radiation.

Spectral profile: A spectral profile is the comparison of the light that is reflected from a sample with the original source signal. Natural and manmade objects have their own spectral signatures that can be used to identify them. The spectral profile charts allow us to select the areas of interest or features on the image and see the spectral information of all bands in a chart format.

Spectral reflectance: Spectral reflectance is the ratio of the energy that is reflected by a surface to the energy that hits the surface at different wavelengths. For example, fresh snow has a very high spectral reflectance of about 80%, while vegetation or forest has a low spectral reflectance of about 5–10%. The value of reflectance also depends on the angle of incidence. It is low at low angles and high at high angles.

$$\text{Spectral reflectance} = \frac{\text{Energy of wavelength reflected from the object}}{\text{Energy of wavelength incident on the object}} \times 100$$

Concepts

Spectral Reflectance Curve

A spectral signature or a spectral response curve is a graph that shows how a surface reflects or emits energy at different wavelengths of the EM spectrum. It helps us understand the characteristics of the surface and how it interacts with the energy. It also helps us choose the best wavelength band for remote sensing data collection. The graph has wavelengths of EM spectrum on x axis and reflectance (%) on y axis. The higher the reflectance, the higher the peak on the graph, which means the surface reflects more energy. The lower the reflectance, the lower the dip on the graph, which means the surface absorbs more energy.

Spectral Reflectance Curve for Vegetation

The reflectance of vegetation depends on the features of individual plants, structure and texture of the canopy, and type and stage of growth of the plants. The reflectance of vegetation is also influenced by internal and external factors that affect the health and maturity of the plants. The reflectance of leaves can be affected by leaf structure, pigmentation, sun exposure, leaf arrangement, hairiness, water content, nutrition, disease, and more. As shown in Figure 1, healthy vegetation absorbs a lot of energy in the visible region (0.4–0.7 μm) because the green chlorophyll pigment in the cells absorbs energy in this

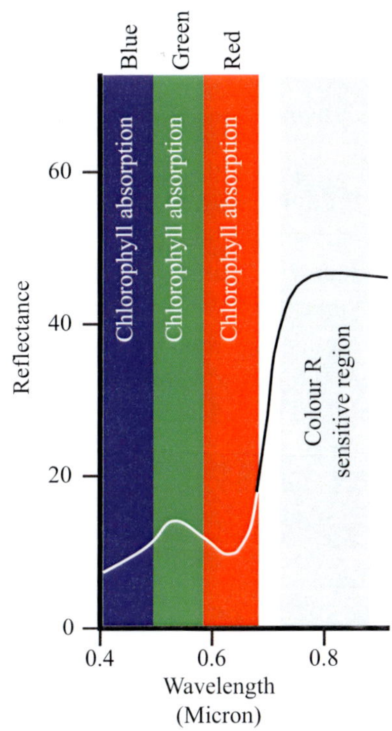

Fig. 1: Spectral reflectance curve for vegetation. (IR: infrared)

region. The reflectance of green light is high, so we see healthy vegetation as green. The absorption decreases sharply at the boundary between red and infrared near 0.7 μm and stays low from 0.7 to 1.3 μm. Then it decreases further for longer wavelengths. The thickness of the canopy also affects the reflectance of infrared light. Thicker canopies reflect more infrared light than thinner ones. This means that we can tell how dense the vegetation is from the reflected infrared light. Stressed or dead vegetation looks yellow or brown because the infrared light is reflected by the dead cells instead of the living cells. This makes the image look darker in this region.

Spectral Reflectance of Soil

Water absorbs energy at 1.4, 1.9, and 2.1 μm wavelengths. Soil reflectance depends on how much water it has, how fine or coarse it is, how rough or smooth it is, how much iron oxide and organic matter it has. Soil with more water reflects less energy, while soil with less water reflects more energy. Sandy soils have less water and higher reflectance, and they look darker than fine soils. Soil with more roughness, organic matter, or iron oxide also reflects less energy.

Spectral Reflectance for Water

Water has different colors depending on the wavelength of the light and the properties of the water. Pure water scatters more light at the blue end of the spectrum, which are the shorter wavelengths. That is why water bodies look deep blue or blue-green. The way water reflects or absorbs light changes with the wavelength and the state of the water. Water in solid form reflects more light at all visible wavelengths than water in liquid form. Water in liquid form reflects more light around 0.6 μm wavelength, but absorbs most of the light beyond 0.7 μm wavelength. The reflectance of water also depends on other factors, such as how much suspended solids, algae, minerals, or particles are in the water. These factors affect how much light is reflected or absorbed by the water.

Spectral Response Patterns

A spectral response pattern is the amount of energy that an object reflects or emits at different wavelengths. The earth's land surface reflects only 3% of the solar radiation that reaches it, while the rest is either reflected by the atmosphere or absorbed and reemitted as infrared energy. Vegetation reflects a lot of the near-infrared (NIR) radiation, which shows that it is healthy and hydrated. Soil that is bare or brownish-gray reflects more energy at longer wavelengths than soil that is covered with grass. Water absorbs most of the radiation at all wavelengths. By looking at the spectral response patterns, we can identify and evaluate different types of crops and geological formations.

Spectral Resolution

Spectral resolution is the ability of a sensor to measure the energy at different wavelengths, in more and narrower bands. Sensors that have 3–10 bands are called multispectral sensors. Sensors that have hundreds or thousands of bands are called hyperspectral sensors. The smaller the wavelength range for a band, the higher the spectral resolution. Spectral resolution helps to distinguish different features and details in an image based on their response to different wavelengths. For example, water and vegetation can be separated by using broad wavelength ranges, such as the visible and NIR. But different rock types may need finer wavelength ranges to be distinguished, so they need a sensor with higher spectral resolution. Spectral resolution is the ability of a sensor to detect fine differences in wavelengths. The higher the spectral resolution, the smaller the wavelength range for a band.

Methodology for Generation of Spectral Profiles

Multispectral imagery is a powerful tool that can distinguish different types of materials. This is because different materials have their own spectral signatures, which are the patterns of reflection and emission of energy at different wavelengths. Spectral signatures can be used to identify materials quantitatively. Spectral profile charts are graphs that show the spectral signatures of different areas or features on an image. They show the spectral information of all bands in a chart format. A spectral profile is the comparison of the light that is reflected from a sample with the original source signal. Natural and man-made objects have their own spectral signatures that can be used to identify them. To create a spectral profile, you can select an area of interest or a feature on the image and enable it. Then, you can choose the plot type that you want to use, such as mean line, box, box and mean line, or consolidated boxes and mean lines. The plot will show the spectral profile data for the selected area or feature. Spectral profile charts have an x axis and a y axis. The x axis shows the number of bands in the image and the wavelengths associated with each band. The y axis shows the pixel values for each band in the image. You can also change the number format of the y axis to display different numeric values. You can also select outliers in the chart, which are pixels that have unusual values. These pixels will be highlighted on the image, so you can see where they are located.

Tools Used in Generation of Spectral Profiles of Different Objects

Satellites

Satellite imagery can help farmers to track and manage their crops under different conditions. It can also help them to estimate the rainfall and plan irrigation and planting times. It can also help them to predict the agricultural output of a region. Some examples of satellites that are used for this purpose include LANDSAT-MSS, LANDSAT-TM, SPOT, and NOAA. These satellites have different sensors and resolutions that can capture different aspects of the crops and the environment. The features of Landsat Multispectral Scanner (MSS) are shown in Table 1.

Sensors

- *High resolution visible (HRV) and high resolution visible infrared (HRVIR) detectors*: HRVIR and HRV are similar sensors, but HRVIR has an extra band that can detect short wave infrared (SWIR) light. Also, the panchromatic mode of HRVIR has a smaller wavelength range than

Table 1: Features of Landsat Multispectral Scanner (MSS).

Landsat MSS 1, 2, 3 spectral bands	Landsat MSS 4, 5 spectral bands	Wavelength	Useful for mapping
Bands 1–3	–	–	Rarely used
Band 4—green	Band 1—green	0.5–0.6	Sediment-laden water, delineates areas of shallow water
Band 5—red	Band 2—red	0.6–0.7	Cultural features
Band 6—Near-infrared	Band 3—Near-infrared	0.7–0.8	Vegetation boundary between land and water, and landforms
Band 7—Near-infrared	Band 4—Near-infrared	0.8–1.1	Penetrates atmospheric haze best, emphasizes vegetation, boundary between land and water, and landforms
Band 8—Thermal infrared		10.4–12.6	Landsat 3 only

the panchromatic mode of HRV. Both HRV and HRVIR can view the earth at an angle, which allows them to take stereo images and to revisit the same place more often, from 1 to 3 days

- *AIRCRAFT-Multispectral Scanner, Multispectral Photograph*: An MSS is a type of scanning system that collects data in different wavelength ranges of the EM spectrum. It can create a two-dimensional image of the surface that shows more details than the human eye can see. It was first developed for military purposes, such as identifying and spying on targets. Later, it was used for space-based imaging platforms to map the earth's features, such as coastlines, vegetation, and landforms. Multispectral imaging can also be used for analyzing documents and paintings. Multispectral imaging uses a small number of spectral bands (usually 3–15), while hyperspectral imaging uses hundreds of continuous spectral bands
- *Spectral bands*: Bands are the specific ranges of wavelengths that a sensor can measure and detect. Different bands can show different details about the surface that is observed. For example, some bands can show vegetation, water, soil, or fire. Here are some categories of bands and their uses:
    - Blue, 450–520 nm, can image the atmosphere and deep water, up to 50 m deep in clear water
    - Green, 520–600 nm, can image vegetation and deep water structures, up to 30 m deep in clear water
    - Red, 630–690 nm, can image man-made objects, soil, and vegetation, up to 9 m deep in water

    - NIR, 750–900 nm, is mainly used for imaging vegetation
    - Mid-infrared (MIR), 1,550–1,750 nm, can image vegetation, soil moisture, and some forest fires
    - Far-infrared (FIR), 2,080–2,350 nm, can image soil, moisture, geological features, silicates, clays, and fires
    - Thermal infrared, 10,400–12,500 nm, uses the radiation that is emitted by the surface instead of the radiation that is reflected by it. It can image geological structures, thermal differences in water currents, fires, and night studies
    - Radar and related technologies can map the terrain and detect various objects
- *Ground—Handheld Radiometers*: Handheld radiometers are small devices that can measure the radiation reflected or emitted by a target. They can be used to collect frequent and detailed data on the spectral and thermal properties of different small plots with various treatments, such as irrigation or fertilization. These experiments help to establish the links between remote sensing data and agricultural variables as well as to improve the interpretation and application of satellite data for agriculture

Advantages of Generation of Spectral Profiles of Different Objects

- With spectral profile charts, you can see the spectral information of all bands for any image area or feature that you pick. A spectral profile has a shape that shows the pixels to choose and an image with key metadata to sample from
- Multispectral imagery can help you tell apart different materials and features in the landscape because they often have distinct spectral signatures that can be used to quantify them
- Spectral signatures measure how different materials reflect or absorb different wavelengths of light. We can use these measurements to get information about an object or region

Disadvantages of Generation of Spectral Profiles of Objects

- Spectral signature is affected by soil variations
- Surface roughness affects the reflection directions of incident light, while soil texture and soil mechanical composition affects the volumetric radiance scattering of incident line
- Spectral profile is affected by the size of viewed area, orientation, and inclination of the view axis, sun elevation, and wind speed

Applications of Spectral Profiles in Agriculture

Table 2: Role of spectral profiles in agriculture

S. No.	Advances	Reference
1.	Sun et al., conducted an experiment using spectral reflectance to estimate leaf chlorophyll content maize inoculated with arbuscular mycorrhizal fungi under water stress.	(Sun et al., 2021)
2.	Saleem et al., used the methodology of the back propagation (BP) approach to determine plant diseases, and this has potential in the analysis of hyperspectral reflectance data.	(Saleem et al., 2019)
3.	Yunus et al., performed a study aimed to apply near-infrared spectroscopy (NIRS) in determining soil macro nutrients, namely, N, P, and K. Based on obtained results, it may conclude that NIRS can applied as an alternative rapid and simultaneous method in predicting soil quality indices.	(Yunus et al., 2019)
4.	Weksler et al., conducted study focused on the potential to identify root damage using high and multispectral resolution during the early stages of a growing season. It was the first to bring proof of concept and demonstrate early-stage spectral reflectance changes caused by root malfunctioning.	(Weksler et al., 2021)
5.	Adak et al., conducted a study to compare the performance of 38 structural spectral reflectance index-based regression models and identify the best model for prediction of grain and biomass yield of wheat under different tillage, residue, and nitrogen management practices. The field experiment was carried out during rabi season of 2016–17 and 2017–18 at ICAR (Indian Council of Agricultural Research)—Indian Agricultural Research Institute, New Delhi. The reflectance of wheat canopy was captured in the spectral range 350–1,800 nm with a bandwidth of 1 nm. Correlation between spectral reflectance indices and grain and biomass yield of wheat indicated that among the booting, milk and dough stages, all the spectral indices at booting stage showed highest correlation with the grain and biomass yield of wheat.	(Adak et al., 2021)
6.	Vibhute et al., conducted experiments in soils of Phulambri Tehsil of Aurangabad District, Maharashtra India using reflectance spectra of soil samples that were obtained using ASD field-spec-4 spectroradiometer and demonstrated that the use of PLSR (partial least squares regression) along with back propagation neural network (BPNN) and genetic algorithm (GA)–BPNN was sufficient for the prediction of soil nitrogen.	(Vibhute et al., 2020)
7.	Srivastava et al., performed the study on visible near-infrared reflectance spectroscopy for rapid characterization of salt-affected soil in the Indogangetic plains of Haryana, India.	(Srivastava et al., 2017)

Table 2: *(Contd...)*

S. No.	Advances	Reference
8.	Guo et al., performed an experiment using a portable spectrometer (ASD, USA) to collect canopy spectral reflectance of soyabean crop. Then leaves ground into fine powder for mineral content analysis.	(Guo et al., 2017)
9.	Panigrahi et al., in their study, evolved a new approach for estimating SWP from spectral reflectance data of a standing rice crop over the visible to shortwave-infrared region (wavelength: 350–2,500 nm).	(Panigrahi and Das 2021)
10.	Ashourloo et al., conducted the experiment with the two most common cultivars of wheat called Bolani and Roshan. A spectroradiometer (Analytical Spectral Device, Boulder, CO, USA) with a 25° field of view was used. The results showed that the disease symptoms have high impact on the infected plant spectra and proposed two indices (LRDSI 1 and LRDSI 2) that not only have high accuracy but also reduce data dimensionality and speed up the disease estimation rate.	(Ashourloo et al., 2014)
11.	Sowmya et al., conducted experiment for red soil in disturbed condition at different stages of moisture content dry to fully saturation by supplying artificial water. The soil reflectance tends to increase with increase in wavelengths from 400 to 1,000 nm because of the presence of the iron oxide absorption at shorter wavelengths. In case of red soil, the portion of spectral reflectance are changes (dips) at 760 nm and 950 nm, which shows the presence of water, and the increase in reflectance after 400 nm which indicates the iron oxide.	(Puram et al., 2019)
12.	Meireles et al., studied how spectra vary across the tree of life or how the evolution of leaf traits affects the differentiation of spectra among species and lineages. Leaf spectra capture the phylogenetic history of seed plants and the evolutionary dynamics of leaf chemistry and structure.	(Meireles et al., 2020)

Conclusion

Remote sensing for precision agriculture applications is a growing field of interest. To use spectral profile imaging to extract accurate and relevant information for agriculture, we need to understand the spectral signatures of different crop types and features. Spectral signatures show the position, shape, variability and similarity of feature-specific absorption bands in the spectrum. This can help us to identify and quantify different agricultural features, such as vegetation, soil types, water areas, crop species or weeds, based on their unique spectral signatures, and their different spectral reflectance measured by various multisensors or hyperspectral sensors.

References

Adak, S., Bandyopadhyay, K.K., Sahoo, R.N., Mridha, N., Shrivastava, M., Purakayastha, T.J. "Prediction of Wheat Yield Using Spectral Reflectance Indices Under Different Tillage, Residue and Nitrogen Management Practices." *Current Science*, 121, 3 (2021): 402.

Ashourloo, D., Mobasheri, M., Huete, A. "Developing Two Spectral Disease Indices for Detection of Wheat Leaf Rust (Pucciniatriticina)." *Remote Sensing*, 6, 6 (2014): 4723-40.

Guo, R., Zhao, M.Z., Yang, Z.X., Wang, G.J., Yin, H., Li, J.D. "Simulation of soybean canopy nutrient contents by hyperspectral remote sensing." *Applied Ecology and Environmental Research* 15, 4 (2017): 1185-98.

Meireles, J.E., Cavender-Bares, J., Townsend, P.A., Ustin, S., Gamon, J.A., Schweiger, A.K., et al. "Leaf reflectance spectra capture the evolutionary history of seed plants." *New Phytologist* 228, 2 (2020): 485-93.

Panigrahi, N., Das, B.S. "Evaluation of regression algorithms for estimating leaf area index and canopy water content from water stressed rice canopy reflectance." *Information Processing in Agriculture* 8, 2 (2021): 284-98.

Puram, S., Mittapalli, G., Prabhakar, M. "Study of Spectral Reflectance Pattern of Red Soils under varying Moisture Conditions." *International Journal for Research in Applied Science & Engineering Technology* 7, IX (2019).

Saleem, P., Mahmood, A. "Plant Disease Detection and Classification by Deep Learning." *Plants* 8, 11 (2019): 468.

Srivastava, R., Sethi, M., Yadav, R.K., Bundela, D.S., Singh, M., Chattaraj, S., et al. "Visible-Near Infrared Reflectance Spectroscopy for Rapid Characterization of Salt-Affected Soil in the Indo-Gangetic Plains of Haryana, India." *Journal of the Indian Society of Remote Sensing* 45, 2 (2017): 307-15.

Sun, J., Yang, L., Yang, X., Wei, J., Li, L., Guo, E., et al. "Using Spectral Reflectance to Estimate the Leaf Chlorophyll Content of Maize Inoculated With Arbuscular Mycorrhizal Fungi Under Water Stress." *Frontiers in Plant Science* 12 (2021).

Vibhute, A.D., Kale, K.V., Gaikwad, S.V., Dhumal, R.K. "Estimation of soil nitrogen in agricultural regions by VNIR reflectance spectroscopy." *SN Applied Sciences* 2, 9 (2020): 1523.

Weksler, S., Rozenstein, O., Haish, N., Moshelion, M., Wallach, R., Ben-Dor, E. Pepper Plants Leaf Spectral Reflectance Changes as a Result of Root Rot Damage. *Remote Sensing* 13, 5 (2021): 980.

Yunus, Y., Devianti, Satriyo, P., Munawar, A.A. "Rapid Prediction of Soil Quality Indices Using Near Infrared Spectroscopy." *IOP Conference Series: Earth and Environmental Science* 365, 1 (2019):012043.

5

Thematic Layers

Abstract

To understand the soil properties better, thematic layers of soil fertility are created. These layers are based on various parameters that affect the soil fertility, such as pH, electrical conductivity, soil structure, soil texture, macro and micro nutrients, toxic elements, and microclimate. These parameters are mapped individually and then combined into a single map with separate layers for each parameter. This way, the user can access all the information in one map and study it more conveniently. The process of creating these maps involves several steps, such as soil sample collection, testing, and mapping. The mapping steps include generating a base map, preparing a boundary of the study area, preparing a drainage map, generating a digital elevation model (DEM), generating a contour map, generating a slope map, generating a soil nutrient availability map, and finally classifying the land cover. These maps are called thematic maps because they focus on one parameter at a time. These maps are then used to create thematic layers. The main purpose of these layers is to provide specific information about a particular location, general information about spatial patterns, and comparison of different parameters. These layers are useful for meeting the increasing demand of site-specific treatments in agriculture, as they help in achieving sustainability and precision farming. They also help in identifying the problems and understanding the soil fertility better, which leads to better crop selection and management practices. These layers consist of raster data, dot data, line data, and polygon data. The raster data is mainly derived from satellite or aerial images, while the dot data, line data, and polygon data are zero-dimensional, one-dimensional, and two-dimensional, respectively. Thematic layers have a huge impact on the agriculture sector as they enhance the soil fertility management.

Keywords: Thematic layers, thematic maps, raster data, point data, polygon data, line data

Introduction

A thematic layer is defined as a layer with a set of logical feature data placed under single theme such as soil type, pH level, salinity, and water table. The features in these layers determine real world objects and are created using geometries such as points, lines and polygons (Tickoo 2017). Thematic layering is the spatial representation of analyzed data of elements that belong to the same type. Thematic maps are used to create these layers, which show how one or more specific data themes vary across different geographic areas. Thematic maps are maps that show the distribution of qualitative or quantitative phenomena of various kinds (Lambert and Zanin 2020). The prime motives of thematic maps is to fulfil the following purposes:

- They provide specific information about the particular locations
- They provide general information about the spatial patterns
- They can be used to compare the patterns on two or more maps

Whereas the goals of the thematic layering are listed as follows:

- The major objective of creating thematic layers of soil fertility is to have better understanding of the soil
- To distinguish the needs of the different soils of different or the same locations
- To have easier and more convenient access to the various parameters that influence the soil fertility
- In the adoption of precision farming
- For comparison of two or more parameters at the same time

The creation of thematic layers of soil fertility is important for the increasing demand of sustainable approaches in agriculture sector. The reasons for this are as follows:

- *Sustainable agriculture is the future*: Sustainability requires using fertilizers and other inputs wisely, based on the soil's needs
- *Proper monitoring*: Thematic layers of soil fertility can help monitor the soil's condition and compare different parameters at the same time, which can improve the yield
- *Increasing soil degradation*: Overuse of chemical fertilizers can degrade the soil quality, so thematic maps and layers can help determine the optimal amount of fertilizers to use

- *Advances in precision farming*: Precision farming depends on knowing the exact needs of the fields, which can be achieved by creating thematic layers of soil fertility
- *Miscellaneous reasons*: These layers also help in understanding the soil fertility and various parameters better, which can be useful for experiments or studies in that area. They also make it easier to study the soil fertility with multiple parameters available at once (Maxwell 2015)

A geographic information system (GIS) represents spatial elements and phenomena as themes or thematic layers. Each layer has only one type of feature data (points, lines, or polygons), or it is an image (raster data model) (Dempsey 2022).

- *Raster data*: Raster data is visual data that is captured by satellites or drones. It shows geographic data as a grid of cells each of which has a value. The area that each cell covers in the real world is called the spatial resolution
- *Point data*: Point data is used to show features that are not connected and to represent discrete data points. These have no dimensions and can be used to mark locations and so on
- *Line data*: Line data is used to show linear features. These have one dimension and can be used to mark streets, rivers, and so on
- *Polygon data*: Polygon data is used to show geographic areas. These have two dimensions and can be used to mark boundaries of enclosed areas, water bodies, and so on

Methodology

Thematic layers are created by using GIS and following several steps. ArcGIS is a software that is used for thematic layering. The first step is to make thematic maps that show one parameter at a time for the land. These maps are then combined to create thematic layers (Sudhakar and Alivelamma 2018) (Yadav et al., 2020). Following steps illustrate how to create a thematic map for soil fertility:

Step 1: Collection of Data

The first step in creating thematic layers is to collect data. This requires testing the soil of the area for elements, like nitrogen, phosphorus, and potassium, to determine its fertility. The data collection process consists of the following steps:

- *Soil sampling*: This is the collection of soil from the area to be studied. GIS can help in identifying the locations for sampling. The sampling methods can be random, grid, zig-zag, intensive, or localized. The soil sample should represent the area well, so the sampling procedure affects the results a lot. The area should be divided into 15–20 sampling units and samples should be taken from a depth of 5–90 cm depending on the crop
- *Soil analysis*: This involves testing the soil samples for properties like pH, electrical conductivity, soil structure, soil texture, macro and micro nutrients, toxic elements, microclimate, etc. This step helps to categorize the samples and the area based on the test results. The analysis should be done carefully for better quality maps and results
- *Classification of area based on test results*: Based on the results, the area should be classified into zones. For example, for nutrient analysis, the zones can be very low, low, moderate, high, very high, etc. For pH, the zones can be acidic, alkaline, or neutral. Similarly, the area can be classified based on other parameters that indicate the soil fertility

Step 2: Preparation of Thematic Map

Thematic maps portray a particular subject matter about a given geographical region, and these are prepared as follows:

- *Generation of base map*: Base map is the framework on which all the data would be entered. This base map can be taken from the state or central institutes. This map has all the simple land features
- *Preparation of boundary of area under study*: In this step the study area is basically delineated on the base map. Here the exact location of the area and its boundary is located on the base map, for this global positioning system (GPS) can be utilized
- *Preparation of drainage map*: In this step the various streams flowing through the area or any other water bodies are located on the map. For field conditions, the irrigation channels are also marked on the base map
- *Generation of DEM*: DEM stands for digital elevation model and it shows the spatial variation in the altitude, i.e., it is the digital representation of

the topographical surface. It can be prepared by using information from GPS using contour lines and elevation points

- *Generation of contour map*: Contour maps are prepared by using GIS technology by utilizing the DEM prepared earlier
- *Generation of slope map*: The slope values can either be measured in degrees or percentage. The slope information is useful in understanding the topography, geomorphology soil types and their erodibility, and surface drainage. Slope is a vital for land capability assessment, formulating soil and water conservation measure. The slope maps are generated using DEM and contour maps
- *Soil nutrient availability map*: The data collected in the step one is used in here. The areas earlier classified based on nutrients are now located on the base map. This gives a general idea of the availability of nutrients in the area under study and hence the fertility of the area
- *Land cover classification*: The land under study is classified according to vegetation grown on it. For this false color composites are utilized

Step 3: Creation of Thematic Layers

As described in the above steps the thematic map is prepared for one nutrient at a time; in a similar manner thematic maps are prepared for other essential nutrients too and other parameters like texture of soil or pH. Then all these are combined as thematic layers, for this any good GIS software can be utilized, e.g., ArcGIS, etc. Using simple steps, the software can be used to generate thematic layers by using the data entered (Yadav et al., 2020).

Scope of Opportunities

- India being a diverse country in terms of types of soil provides a vast scope for the study of soil fertility
- Indian soils show severe toxicity of some elements and severe deficiency of others, making it a great subject of study, hence providing huge opportunities for the development of thematic layers of soil fertility
- Land resources can be classified easily once mapping is done
- Precision farming is growing rapidly in India, making a vast scope of thematic layering
- Indian agriculture is shifting toward sustainability slowly, hence increasing the demand of thematic layers of soil fertility

Challenges

- The most important challenge is the lack of proper data required for the preparation of these layers
- The thematic layers require GIS software, which though available in the country are not as commonly available as it seems
- The process of creation of thematic layers is complex and requires highly specialized skill set
- These layers even when available are used only by the researchers and not the farmers for whom these are made
- There is still a lack of awareness about these resources among the farmers

Impacts

Creation of thematic layers of soil fertility have proven to be a great help to agriculture sector. The major impacts are listed below:

- *Better understanding of needs of crop*: With the development of thematic layers of soil fertility, farmers can have the exact requirements of their fields, hence their crops; this enables them to know the amount of fertilizers, etc. required and the time of their application
- *Knowledge of what practices to use*: These maps provide information of the condition of the soil, hence enabling the farmers to choose better practices for better yields
- *Identification of problems*: These provide information about the problems of the area and the remedies to be applied in order to get better yields
- *Better information to researchers*: These thematic maps provide exact information to the researchers, hence providing better information of geographic conditions along with the knowledge of problems and requirements of the area, this opens up gates to better research hence better knowledge of the area. This helps in better innovations leading to improved agricultural practices
- Thematic layering enables the reader to work on several parameters at the same time, hence providing vast scope of modifications and improvements in existing agricultural practices
- Thematic layering gives the exact condition of soil fertility in a given area considering various parameters at the same time; this helps the farmer in choosing the most beneficial crop for the season

Advances Till Date

In Russia, a detailed study was performed to know the various reasons for low fertility of agricultural lands, for this all the parameters governing the fertility of a land were monitored like the micro- and macronutrient content in the soil, soil pH, soil salinity, level of water, electrical conductivity, cation exchange capacity, anion exchange capacity, and several other such factors. Other factors that affect land productivity were also studied like proper management practices. Based on the results as such the maps were produced and later on better management practices were developed based on the results of the map as produced (Klyushin et al., 2017). In Hassan district of Karnataka, a detailed study was conducted in 2019 on Honnavalli watershed to know the soil fertility status of the region by integration of both GIS and GPS techniques and as such fertility maps were created using the information so collected. As per the study of the watershed, samples were taken from an area of 420 ha each at 400 m intervals and analyzed for surface soil physicochemical and biological properties. The results revealed that watershed found acidic pH area (93%), low OC (55% area) and normal electrical conductivity (100% area), further the watershed area was low in available N (51% area), P_2O_5 (77% area), and medium in K_2O (48% area). The exchangeable Ca and Mg were sufficient in 95% and 72% of the watershed area, respectively. The 95% of the watershed area has sufficient available Sr. The surface soil fertility maps of pH, electrical conductivity, organic carbon, available N, P and K, exchangeable Ca and Mg and available S were prepared through ArcGIS (Biradar et al., 2020).

Merits and Demerits

The various advantages of thematic layering of soil fertility are listed below:

- Data is easily available
- Farmers can easily identify the issues with the area and rectify accordingly
- Several parameters can be studied at the same time reducing the need to look at several different thematic maps, hence, time-efficient
- It provides site-specific requirements of soil hence making agricultural practices sustainable
- Thematic layers of soil fertility can also help in yield estimation beforehand
- These make the study of soil fertility of given area a lot easier for researchers

- Thematic maps can be used for spatial data analysis, formulating hypotheses, synthesizing spatial data by revealing patterns and relationships, and data presentation

Conclusion

Thematic layers have some drawbacks, but their benefits are more important. They can be very useful for the farmers if they are trained and informed properly. They can help achieve sustainability and profitability in the agriculture sector. Thus thematic layers of soil fertility should be created to make the best use of this natural resource, the soil.

References

Biradar, B., HM, J., S, C., KN, G., Sannagoudar, M.S., AS, P., et al. "Assessment of soil fertility through GIS techniques and thematic mapping in micro-watershed of Hassan, Karnataka." *Journal of Pharmacognosy and Phytochemistry* 9, 4 (2020): 3218-28.

Dempsey, C. 2022. Types of GIS Data Explored: Vector and Raster https://www.gislounge.com/geodatabases-explored-vector-and-raster-data/ [Last accessed February, 2024].

Klyushin, P.V., Volkov, S.N., Shapovalov, D.A., Shirokova, V.A., Khutorova, A.O. "Solutions of problems in defining indicators of agricultural land within the framework of activities for the implementation of the concept of development monitoring in the Russian Federation." *Section Ecology and Environmental Protection* 17 (2017): 819.

Lambert, N., Zanin, C. 2020. *Practical Handbook of Thematic Cartography. Principles, Methods, and Applications*. United States: CRC Press.

Maxwell, R. 2015. Maps Serve to Monitor Global Agriculture. https://www.gislounge.com/maps-serve-to-monitor-global-agriculture/ [Last accessed February, 2024].

Sudhakar U., Alivelamma K. "Original Research Paper Geography Generation of Thematic Layers in Mothkuri Watershed Using in Geo-spatial Techniques Nalgonda District." *Indian Journal of Applied Research* 12 (2018): 54-6.

Tickoo, S. 2017. *Exploring AUTOCAD MAP 3D 2018*. United States: Cadcim Technologies.

Yadav, S.S., Mishra, V., Kumar, V., Sahu, S., Kumar, S., Katre, P. "Preparation of Thematic Maps and Farming Situation Characterisation of Distributary Command Area using RS and GIS Techniques." *International Journal of Current Microbiology and Applied Sciences* 9, 2 (2020): 992-1000.

6

Crop Differentiation, Yield Monitoring, and Soil Mapping

Abstract

Crop differentiation is the process of identifying crops based on their season, growth stage, structure, and spectral characteristics. It helps to distinguish different types of crops from each other. Crop differentiation is followed by yield monitoring and yield mapping, which measure and visualize the crop production in a given area. These information can be used to create a map that shows different zones or ranges of crop performance using different colors. This map can help policy makers to decide which interventions are needed to improve crop sustainability.

Keywords: Crop discrimination, GLCM, remote-sensing, intervention

Introduction

Crop discrimination is the process of using remote-sensing technologies to identify and characterize different types of crops based on their electromagnetic behavior at different wavelengths. Each crop has a unique spectral signature that reflects its response to the incident radiation. However, the spectral signature can also vary depending on the season, the angle of the sensor, crop's features, illumination intensity, weather phenomenon, and topography among other external factors (Parida et al., 2023). After discriminating the crops, the next step is to monitor and map their yield using farm equipment such as drones, tractors, or harvesters that can collect information such as grain yield, moisture levels, and soil properties during the harvesting period (Wei et al., 2023). The information is then displayed on a screen and used to create a map that shows the crop performance in different zones or ranges. This map can help in making better decisions for improving crop productivity and sustainability (Li and Zhang, 2023; Roy et al., 2010).

Tools and Techniques

It uses techniques of remote-sensing by understanding about electromagnetic behavior of the item or object.

Remote-sensing

- Different crops have different growth phases, characteristics, or seasonal patterns, so their growth rates vary at different times. Using these crop growth characteristics to enhance the crop growth difference information at different times is essential for early crop identification. This principle is applied by remote sensing, which captures large area images using sensor technologies and spectral resolutions imageries and nonimaging spectroradiometer. These imaging and nonimaging data are used to characterize the different crop species. The same crops in the same growing season can have different durations and magnitudes of each growth phase, which can create data variability for crop type discrimination. Hyperspectral data are used to characterize, classify, model, and map agricultural crops (Cherlinka 2023)

Remote-sensing Indices

- They are better characterized by vegetation indices (VIs) than by spectral bands (MINNI 2023). VIs are combinations of surface reflectance at two or more wavelengths designed to highlight a particular property of vegetation. They are derived from the reflectance properties of vegetation. Each of the VIs is designed to emphasize a specific vegetation property. Some of the commonly used VIs are Ratio Vegetation Index (RVI), Normalized Difference Vegetation Index (NDVI), Vegetation Condition Index (VCI), Temperature Condition Index (TCI), Temperature–Vegetation Dryness Index (TVDI), Relative Greenness (RG), Vegetation Temperature Condition Index (VTCI), Enhanced Vegetation Index (EVI), NDVI Deviation of long-term mean (DEV), Ratio Vegetation Index (RVI), Ratio Drought Index (RDI), Rapid Indicator of the Drought (RID), Vegetation-Temperature Index (VTI), Difference Vegetation Index (DVI), Transformed Normalized Vegetation Index (TNDVI), Perpendicular Vegetation Index (PVI) and Greenness vegetation index (GVI) (Auravant 2022)

Feature Extraction

Image processing techniques in agriculture are often used to segment the background, eliminate the crop, and extract the plant from the field images. This helps to detect and classify different types of plants and identify any problems.

To improve the accuracy of detection, morphological operations can be performed to extract different patterns from the crop/weed leaf images (Cunha et al., 2023). Image contour features are also useful for plant classification and localization, as they capture the shape and boundary information of the plant (Bouguettaya et al., 2022; Kitzler et al., 2023). Feature extraction is the process of defining image characteristics or features that provide meaningful information for image interpretation or classification (Wang et al., 2021).

Role of Texture in Classification

Unmanned aerial vehicle (UAV) imaging is a promising technique for obtaining high-resolution and timely aerial photos of crops, which can overcome the cost and environmental limitations of conventional methods. However, pixel-based analysis of UAV images may not be sufficient to capture the spatial patterns and variations of different crops. Therefore, texture features can be extracted from the gray-level co-occurrence matrix (GLCM) to enhance the crop classification accuracy. Texture refers to the spatial arrangement of colors or intensities in an image, which can distinguish man-made objects with regular textures from natural objects with irregular textures. GLCM, developed by Haralick *et al.* (Haralick et al., 1973), is a widely used statistical method for extracting texture features from remote-sensing data for vegetation structure modeling and land cover classification (Liu et al., 2021; Velusamy et al., 2021). GLCM is based on the frequency of occurrence of pairs of gray-level pixels in a given spatial relationship. To construct a GLCM, the original image is first converted to gray-scale, and then the frequency of each pair of gray-levels is counted in a specified direction and distance (Shi et al., 2022). The GLCM can be used to calculate various texture measures, such as contrast, homogeneity, energy, and entropy.

Yield monitoring: Yield monitoring is a feature of precision agriculture that helps to provide information to the farmers that is useful for various field management purposes. It involves estimating the amount of nutrients removed by the harvested crop, predicting the grain yield before harvest, measuring the moisture levels, soil properties, calculating the profitability, developing management zones, and analyzing the impacts of treatments that are used in on-farm studies. This helps the farmers to make better decisions about how to sow, what amount of fertilizer to use, when to harvest, and how to cope with the effects of weather (Farms.com 2021; Institute of Agriculture and Natural Resources; CropWatch 2002). Yield monitoring works by measuring the weight and moisture content of the grain as it is harvested using sensors attached to the grain elevator. This data is then georeferenced and recorded using a GPS receiver and a yield monitor display (Jagdish 2023). Remote-sensing data from aerial photography, multispectral scanners, radar, or satellite can also be used to predict crop yields and provide representative and spatial information of crop yield estimation (Grisso et al., 2014). Remote-sensing data can capture the crop growth conditions, VIs, biomass, leaf area index, etc., which are related to crop yield potential (Gumma et al., 2022).

Yield mapping: Yield mapping is a precision agriculture tool that collects georeferenced data on the yield and moisture content of crops in different parts of a field. It uses various sensors mounted on a combine harvester to measure these parameters in real time as the crop is harvested (Institute of Agriculture and Natural Resources; CropWatch 2002). The main components of a grain yield mapping system are (BRUNEL 2016):

- *Grain flow sensor*: Measures the volume of grain harvested
- *Grain moisture sensor*: Adjusts the yield for moisture variability
- *Clean grain elevator speed sensor*: Improves the accuracy of grain flow measurement (optional)
- *GPS antenna*: Receives satellite signals for geolocation
- *Yield monitor display with a GPS receiver*: Records and displays the data
- *Header position sensor*: Detects when the harvester is turning or out of the crop
- *Travel speed sensor*: Measures the distance traveled by the harvester (can be replaced by GPS data)

Each sensor needs to be calibrated properly according to the operator's manual. The data from the sensors are recorded in a binary log file that can be converted to a text format or a map using the yield monitor software. Yield mapping is a powerful tool for precision agriculture, as it can help farmers to analyze the factors affecting the yield, forecast the future needs, and apply variable rate inputs according to spatially variable yield goals (Barocco et al., 2017). Ideally, at least 5 years of yield maps are needed to capture the temporal variability of the yield.

Soil mapping: Remote-sensing technology is a powerful tool for providing reliable spatial information on different natural resources. It uses various space-borne sensors that can capture the earth's surface at different scales and frequencies. It also allows for the comparison of past and present conditions of the resources by using historical data (Kumar et al., 2015). GIS is another tool that enhances the analysis and integration of spatial information for economic development and environmental protection (Nowatzki et al., 2017).

For soil mapping, remote-sensing and GIS can help to identify the landform variations based on topography, slope, aspect, and land cover, which affect the soil properties (Shustova 2022). These tools can also help to delineate the soil boundaries and sampling locations more accurately and efficiently than conventional methods.

Method: Soil mapping using GPS and satellite data is a technique to collect and analyze spatial information on soil properties and variability. It involves the following steps:

- *Soil sampling*: GPS is used to locate and record the soil samples on a grid pattern of 2.5 acres, with 4–6 samples taken in each grid. The soil samples are numbered and georeferenced using GPS, and the attribute data is stored in a database (Woydziak and 2013)
- *Soil analysis*: The soil samples are analyzed for various parameters, such as texture, pH, salinity, organic matter, and nutrients. The results are used to classify and characterize the soil types and zones (Ackerson 2018; Grind GIS 2018)
- *Satellite imagery*: Satellite data is used to capture the reflectance of the soil surface under direct solar radiation. The satellite image should have good contrast to distinguish different land features and soil types. The optimal season for satellite data acquisition varies depending on the region, but generally it is between February and April, when the crops are harvested or not yet sown (Gasmi et al., 2022; Sahu et al., 2021)
- *Image processing*: The satellite image is processed using various techniques, such as enhancement, classification, and segmentation, to identify and delineate the soil boundaries and patterns (Žížala et al., 2019)
- *Map generation*: The satellite image and the soil analysis data are integrated using GIS to generate a soil map that shows the spatial distribution of soil properties and zones. The soil map can help the farm managers to determine the appropriate fertilization applications for each zone (Guo et al., 2022)

Use of scale in soil mapping: The scale of a soil map is related to the amount of information it contains and the intensity of field investigations. Different scales of soil maps are suitable for different purposes and levels of planning. The following are some examples of soil map scales and their applications (FAO 2022):

- *1:1 million scale*: These are small-scale soil maps that provide general information on soil types and regions. They are useful for macro level planning at the national level, such as assessing soil resources and potential land uses
- *1:250,000 scale*: These are medium-scale soil maps that provide more detailed information on soil classes and properties. They are useful for regional or state level planning, such as evaluating soil suitability for various agricultural uses
- *1:50,000 scale*: These are large-scale soil maps that provide information on soil series and subtypes. They are useful for district level planning, such as designing resource conservation and optimal land use strategies
- *1:8,000 or 1:4,000 scale*: These are very large-scale soil maps that provide information on soil characteristics and variability at the field level. They are useful for specific purposes, such as precision agriculture and site-specific management

The scale of soil mapping is also determined by the spatial and spectral resolution of remote-sensing data. Remote-sensing data can capture the reflectance and absorption of the soil surface at different wavelengths, which can indicate the soil properties and patterns (Brewer 2011). The following are some examples of remote-sensing data sources and their corresponding scales of soil mapping (Heil et al., 2022):

- *Coarse resolution data (70 m or higher)*: These data are obtained from sensors such as IRS LISS-I, WIFS, AWIFs, and LANDSAT-MSS. They can be used to prepare soil maps at 1:250,000 scale or smaller, which show the broad soil regions and types
- *Medium resolution data (10–30 m)*: These data are obtained from sensors such as LANDSAT-TM, IRS LISS-II/LISS-III, and SPOT-MLA. They can be used to prepare soil maps at 1:50,000 scale, which show the soil classes and properties
- *High-resolution data (1–5 m)*: These data are obtained from sensors such as IRS-P6 (LISS-IV), Cartosat-1 and Cartosat-2, and IKONOS. They can be used to prepare soil maps at 1:10,000 scale or larger, which show the soil series and subtypes

Image interpretation: Remote-sensing data can be interpreted in two ways: visually or digitally. Visual interpretation is a simple and cheap method that involves examining the images or photographs by human eyes. Digital interpretation is a more complex and expensive method that involves processing the images or data by computer software. For soil mapping, visual interpretation of remote-sensing data is often used as a basic approach. The steps involved in this approach are (e-Krishi Shiksha 2014; Sivakumar 2010):

- *Prefield interpretation*: Satellite data and topographic maps are used to identify and delineate the different physiographic units, such as slope, aspect, and land cover that affect the soil properties
- *Ground truth collection and verification*: Soil profiles are studied and soil samples are collected and analyzed in the laboratory to characterize the soil types and parameters. The physiographic units are verified by comparing them with the ground observations
- *Physiography–soil relationship*: The relationship between the physiographic units and the soil types is established based on the soil profile data and the satellite image interpretation. This relationship is used to classify and map the soils in the study area
- *Extrapolation*: The physiography–soil relationship is extrapolated to other similar areas that have not been sampled or studied, but have similar physiographic features. This helps to extend the soil mapping to larger areas

Conclusion

Crop classification or discrimination is the process of identifying the area and type of different crops or varieties within the same crop. It can help to monitor and evaluate the crop production and diversity. Yield monitoring is another important aspect of precision agriculture, as it helps farmers or policy makers to measure and visualize the yield variability within a field. It can provide valuable information in the form of yield or soil maps that show the high and low production areas of the field. This information can help to make better decisions on tillage methods, fertilization, irrigation, crop rotation, etc., to improve and sustain the crop productivity and profitability.

References

Ackerson, J. 2018. Soil Sampling Guidelines. https://www.extension.purdue.edu/extmedia/AY/AY-368-w.pdf [Last accessed February, 2024].

Barocco, R., Lee, W. S.D., Hortman, G. "Yield Mapping Hardware Components for Grains and Cotton Using On-the-Go Monitoring Systems." *EDIS* 3 (2017).

Bouguettaya, A., Zarzour, H., Kechida, A., Taberkit, A.M. "Deep learning techniques to classify agricultural crops through UAV imagery: a review." *Neural Computing and Applications* 34, 12 (2022): 9511-36.

Brewer, J. 2011. What are Soil Map Units and Web Soil Survey James Brewer Resource Soil Scientist USDA-NRCS 2011 Maryland Soil Survey Work Planning Conference.

Brunel. 2016. Yield Mapping Part 1—What you need to know. https://www.antaraag.ca/precision-agriculture/yield-mapping-101-part-1/ [Last accessed February, 2024].

Cherlinka, V. 2023. Stages of Plant Growth for Improved Crop Management. https://eos.com/blog/stages-of-plant-growth/ [Last accessed February, 2024].

Cunha, N., Barros, T., Reis, M., Marta, T., Premebida, C., Nunes, U.J. 2023. Multispectral Image Segmentation in Agriculture: A Comprehensive Study on Fusion Approaches. https://arxiv.org/abs/2308.00159v1 [Last accessed February, 2024].

e-Krishi Shiksha. 2014. Remote Sensing & GIS Applications 3: Lesson 9 Image Interpretation. 3, 1–8. http://ecoursesonline.iasri.res.in/mod/page/view.php?id=124946 [Last accessed February, 2023].

FAO. 2022. Global Map of Salt-affected Soils. FAO Soils Portal. https://www.fao.org/soils-portal/data-hub/soil-maps-and-databases/faounesco-soil-map-of-the-world/en/ [Last accessed February, 2023].

Farms.com. 2021. Yield Monitoring in Precision Agriculture. https://www.farms.com/precision-agriculture/yield-monitoring/ [Last accessed February, 2023].

Gasmi, A., Gomez, C., Chehbouni, A., Dhiba, D., Elfil, H. "Satellite Multi-Sensor Data Fusion for Soil Clay Mapping Based on the Spectral Index and Spectral Bands Approaches." *Remote Sensing* 14, 5 (2022): 1103.

Grind GIS. 2018. 30 Uses or Applications of GPS in Agriculture. https://grindgis.com/blog/30-uses-of-gps-in-agriculture [Last accessed February, 2023].

Grisso, R.D., Alley, M.M., McClellan, P. Precision Farming Tools. Yield Monitor. https://vtechworks.lib.vt.edu/handle/10919/51375 [Last accessed February, 2024].

Gumma, M.K., Kadiyala, M.D.M., Panjala, P., Ray, S.S., Akuraju, V.R., Dubey, S., et al. "Assimilation of Remote Sensing Data into Crop Growth Model for Yield Estimation: A Case Study from India." *Journal of the Indian Society of Remote Sensing* 50, 2 (2022): 257-70.

Guo, H., Zhang, R., Dai, W., Zhou, X., Zhang, D., Yang, Y., et al. "Mapping Soil Organic Matter Content Based on Feature Band Selection with ZY1-02D Hyperspectral Satellite Data in the Agricultural Region." *Agronomy* 12, 9 (2022): 2111.

Haralick, R.M., Shanmugam, K., Dinstein, I. "Textural Features for Image Classification." *IEEE Transactions on Systems, Man, and Cybernetics* SMC-3, 6 (1973): 610-21.

Heil, J., Jörges, C., Stumpe, B. "Fine-Scale Mapping of Soil Organic Matter in Agricultural Soils Using UAVs and Machine Learning." *Remote Sensing* 14, 14 (2022): 3349.

Jagdish. 2023. Importance of Yield Monitors in Agriculture: Advantages and Working Principle. https://www.agrifarming.in/importance-of-yield-monitors-in-agriculture-advantages-and-working-principle [Last accesssed February, 2024].

Kitzler, F., Barta, N., Neugschwandtner, R.W., Gronauer, A., Motsch, V. "WE3DS: An RGB-D Image Dataset for Semantic Segmentation in Agriculture." *Sensors* 23, 5 (2023): 2713.

Kumar, N., Yamaç, S., Velmurugan, A.V. "Applications of Remote Sensing and GIS in Natural Resource Management." *The Andaman Science Association*, 20 (2015): 1-6.

Li, M., Zhang, M. 2023. "Yield Monitoring and Mapping Technologies." *Encyclopedia of Smart Agriculture Technologies.* United States: Springer International Publishing. pp. 1-6.

Liu, J., Xiang, J., Jin, Y., Liu, R., Yan, J., Wang, L. "Boost Precision Agriculture with Unmanned Aerial Vehicle Remote Sensing and Edge Intelligence: A Survey." *Remote Sensing* 13, 21 (2021): 4387.

Minni, M. (2023). Phases of Growth: Different Phases and Explanation. https://www.embibe.com/exams/phases-of-growth/ [Last accessed February, 2024].

Woydziak, N. 2013. GPS Soil Sampling—The Basics Exposed. CropQuest. https://www.cropquest.com/gps-soil-sampling/ [Last accessed February, 2024].

Nowatzki, J., Andres, R., Kyllo, K. 2017. Agricultural Remote Sensing Basics. https://www.ag.ndsu.edu/publications/crops/agricultural-remote-sensing-basics [Last accessed February, 2024].

Parida, B.R., Kumar, A., Ranjan, A.K. "Crop Types Discrimination and Yield Prediction Using Sentinel-2 Data and AquaCrop Model in Hazaribagh District, Jharkhand." *KN-Journal of Cartography and Geographic Information* 73, 1 (2023): 77-89.

Institute of Agriculture and Natural Resources; CropWatch. 2002. Yield Monitoring and Mapping | *CropWatch*. https://cropwatch.unl.edu/ssm/mapping [Last accessed February, 2024].

Roy, P.S., Dwivedi, R.S., Vijayan, D. 2010. *Remote Sensing Application*. Hyderabad: NRSC, ISRO.

Sahu, K.K., Kar, S., Rout, S. 2021. "Satellite-based soil erosion mapping." *Sustainable Agriculture: Technical Progressions and Transitions*. Switzerland: Springer Nature. pp. 225-52.

Shi, Y., Gao, Y., Wang, Y., Luo, D., Chen, S., Ding, Z., et al. "Using Unmanned Aerial Vehicle-Based Multispectral Image Data to Monitor the Growth of Intercropping Crops in Tea Plantation." *Frontiers in Plant Science* 13 (2022).

Shustova, A. 2022. How earth observation helps natural resource management. https://dragonflyaerospace.com/earth-observation-and-natural-resource-management/ [Last accessed February, 2023].

Sivakumar, R. 2010. Image Interpretation of Remote Sensing data. https://www.geospatialworld.net/article/image-interpretation-of-remote-sensing-data/ [Last accessed February, 2023].

Auravant. Vegetation indices and their interpretation: NDVI, GNDVI, MSAVI2, NDRE, and NDWI - Auravant. 2022. https://www.auravant.com/en/articles/precision-agriculture/vegetation-indices-and-their-interpretation-ndvi-gndvi-msavi2-ndre-and-ndwi/ [Last accessed February, 2023].

Velusamy, P., Rajendran, S., Mahendran, R.K., Naseer, S., Shafiq, M., Choi, J.G. "Unmanned Aerial Vehicles (UAV) in Precision Agriculture: Applications and Challenges." *Energies* 15, 1 (2021): 217.

Wang, S., Sun, G., Zheng, B., Du, Y. "A Crop Image Segmentation and Extraction Algorithm Based on Mask RCNN." *Entropy* 23, 9 (2021): 1160.

Wei, M., Wang, H., Zhang, Y., Li, Q., Du, X., Shi, G., et al. "Investigating the Potential of Crop Discrimination in Early Growing Stage of Change Analysis in Remote Sensing Crop Profiles." *Remote Sensing* 15, 3 (2023).

Žížala, D., Minařík, R., Zádorová, T. "Soil Organic Carbon Mapping Using Multispectral Remote Sensing Data: Prediction Ability of Data with Different Spatial and Spectral Resolutions." *Remote Sensing* 11, 24 (2019): 2947.

7

Precision Agriculture

Abstract

Precision agriculture (PA) aims to adjust the input, such as fertilizer, water, or pesticides, according to the variable temporal and spatial characteristics of the farm. For this purpose, variable rate technologies are used, which are equipment that can automatically change the amount of material usage based on the spatial position of the farm. PA also relies on sensors and converters, which enable the automatic evaluation and guidance of the equipment. One of the important aspects of PA is creating the Yield Map, which shows the variation of yield across the farm. The Yield Map can help to identify the factors and problems that affect the crop production. The Yield Map can be created using Global Positioning System (GPS) with high accuracy and then integrated with local geographic data in geographic information system (GIS) network, which can facilitate the farm management practices.

Keywords: Precision agriculture, global positioning system (GPS), sustainable development, geographic information system (GIS), Yield Map

Introduction

Agriculture is a vital sector in the world economy, as it helps to reduce poverty and promote sustainable development, especially in developing countries. However, agriculture also faces challenges such as increasing productivity and reducing environmental impacts. Therefore, new technologies and methods are needed to improve agricultural practices. One of these methods is precision agriculture (PA), which is based on the principle of sustainable agriculture and healthy food production. PA is a science that uses high technology sensors and analysis tools to improve crop yields and support management decisions. PA is a new concept that aims to increase production, reduce labor time, and optimize the management of fertilizers and irrigation processes. It collects and analyzes a large amount of data and information to enhance the use of agricultural resources, yields, and the quality of crops (Mulla 2013). PA is a new advanced method in which farmers provide optimized inputs such as water and fertilizer to enhance productivity, quality, and yield (Gebbers and Adamchuk 2010). PA is an approach that uses information technology (IT) to optimize the health and productivity of crops. It is also called satellite farming or site-specific crop management (SSCM) because it uses satellite data and Global Positioning System (GPS) to monitor and manage the variability in crop conditions within and between fields. Phytogeomorphology is an aspect of PA that relates the crop growth stability and characteristics to the topographic features of the terrain. This is based on the assumption that the geomorphology influences the hydrology and soil properties of the farm. PA employs sensors, robots, GPS, mapping instruments, and data analysis software to customize the care of plants without increasing labor. Sensors and drones collect images and data on individual plants, such as stem size, leaf shape, and soil moisture, and send them to a computer that analyzes them for signs of health and stress. Farmers receive real-time feedback and can adjust the water, pesticide or fertilizer application to the specific needs of each plant or zone. PA can also help farmers decide when to plant and harvest crops based on the crop growth stage and environmental conditions (Digiteum 2021; Lu et al., 2022; Nowak 2021; Pandya et al., 2021).

Concepts

- *Precision farming takes into consideration the natural resources variability*: This includes soil and landscape variability, digital elevation models, soil mapping, geostatistics, geographic information systems (GIS), microclimate, weather forecasting, remote-sensing, management units, scale, etc. Thus it deals with the measurement and analysis of the spatial and temporal variability of natural resources that affect crop production and management and manage variability through sampling techniques, site-specific nutrient, and crop protection chemical recommendation, crop quality, tillage, seed density, seed variety, yield mapping, remote-sensing, record keeping systems, data interpretation and use, crops (corn, wheat, sugar beets, potatoes, peanut, cotton, vegetables, etc.), management scale, etc. This application of PA technologies and practices help manage the variability of natural resources and optimize the crop performance and profitability. For this management, it may use engineering technologies, viz., computers, positioning systems, DGPS, machinery, tillage, planting, nutrient and crop protection implements, manure, irrigation, fertigation, yield monitor and mapping, soil physical and chemical characteristic sensors, weed/pest mapping, etc., for improvement of PA hardware and software system that enable the collection and processing of data and the implementation of site-specific management actions

- *Profitability*: For the assessment and comparison of the economic and environmental benefits and costs of PA technologies and practices at different scales and contexts the indices include MEY (maximum economic yield), net profit, BMP (best management practices), best recommendations, crop quality, technology costs, sustainability, social impact, marketing, cooperatives, farm size, crop type, etc. Further impact and mitigation of environmental problems PA thrust on the practices of the environment and natural resources quality and conservation

  This may include nutrients, pesticides, sedimentation materials, leaching, runoff, practices, fields, watersheds, on/off-farms, artificial drainage, groundwater, surface water, etc

Methodology of Precision Agriculture

Precision agriculture is a method of managing farms using IT to ensure optimal health and productivity of the crops. It uses GPS, sensors, and other tools to measure and adjust the inputs, such as fertilizer, water, or pesticides, according to the variability in crop conditions within and between fields. One of the steps in PA is to create Yield Maps, which show the variation of yield across the fields. Yield Maps can help to identify the factors and problems that affect the crop production and to apply variable rate inputs according to spatially variable yield goals. To create accurate Yield Maps, the following steps are required (Farmaha et al., 2020; Sinha 2019):

- *Collecting accurate yield data*: This involves collecting soil and yield data using GPS and sensors. Soil samples are taken and analyzed to determine soil texture, organic matter, and nutrient levels. Yield data are collected using sensors installed on the combine harvester that measure the grain flow, moisture content, and harvest speed. These data are georeferenced and recorded using a GPS receiver and a yield monitor display
- *Data storage*: This involves copying the raw data files onto a CD or a computer as a backup. The data files should have consistent and descriptive names that include the field name, crop type, year, and data type (such as raw, trimmed, or interpolated)
- *Data cleaning*: This involves removing the data points that do not represent the true yield at a corresponding location. These data points

can be caused by GPS signal errors, combine operation errors, or sensor calibration errors. Data cleaning can be done using software tools such as Yield Editor or Microsoft Excel. Data cleaning can improve the accuracy and reliability of Yield Maps

Techniques, Technologies, and Equipment Used in Precision Agriculture

Mapping

The creation of maps for crop and soil attributes is the most important and first step in PA. These maps can show the geographic variability of crop and soil properties and help to manage them accordingly. The use of GPS can improve the data collection process by providing accurate location coordinates before and during crop production. Different technologies can be used to collect data, such as grid soil sampling, yield monitoring, remote-sensing, and crop scouting. Data can be collected during crop cultivation using various sensing tools, such as soil probes, electrical conductivity (ECa) meters, and soil nutrient analyzers. Remote-sensing, GIS, and manual mapping during field work are some of the methods that can be used to create maps (FAO-UNESCO 2024; Manupriya 2021).

Global Positioning System Receivers

Global Positioning System is a system of satellites that send signals to GPS receivers on the ground. These receivers can use the signals to calculate their location in real time. This means that they can keep track of their position while moving. This also enables them to map the soil and crop measurements that they collect in the field. GPS receivers can be handheld or mounted on farm equipment. They can help users to locate specific spots in the field where they need to sample or treat the soil or crops.

Sensor Technologies

Measurements of humidity, vegetation, temperature, texture, structure, physical character, conductivity, photoelectricity, and ultrasound are made using a variety of methods, including electromagnetic, conductivity, photoelectricity, and ultrasound. Crop species can be distinguished, stress points can be found, pests and weeds can be identified, and drought, soil, and plant conditions can all be tracked using remote-sensing data (Chen et al., 2004). Sensors allow for the capture of enormous amounts of data without the need for laboratory examination. The use of sensors to provide data on soil characteristics and plant fertility/water status has been widely reported and is a key component of precision agricultural technology. Surveying the field with soil apparent ECa sensors, which continually capture data when moved over the field surface, is one of the most common methods for describing soil variability. Since ECa is sensitive to variations in salinity and soil texture, these sensors offer a great starting point for site-specific control.

Precision Irrigation in Pressurized Systems

Sprinkler irrigation is a type of microirrigation that uses sprinklers to apply water to the crops. Recent innovations in sprinkler irrigation use GPS-based controls to regulate the movement of the irrigation equipment. This can improve the accuracy and efficiency of water application. Moreover, wireless communication and sensor technologies are being developed to monitor the soil and environmental conditions as well as the irrigation machine parameters (such as flow and pressure). This can enhance the water use efficiency and crop productivity. These technologies are very promising, but they need more research and development before they can be commercially available.

Yield Monitoring and Mapping

Grain yield monitors are devices that measure and record the grain flow in the clean grain elevator of a combine harvester. By connecting a yield monitor to a GPS receiver, the grain yield data can be georeferenced and mapped. Yield Maps are useful for evaluating the crop performance and the effects of soil, terrain, and other environmental factors. Yield Maps can also help to assess the impacts of management practices such as seed, herbicide, fertilizer, irrigation, and tillage (Farms.com 2024; Institute of Agriculture and Natural Resources; CropWatch 2002). To create accurate Yield Maps, the yield monitor data need to be calibrated, cleaned, and processed using software tools (Pioneer 2019). Grid soil sampling is a technique that collects soil samples in a regular grid pattern across a field. Each soil sample has a GPS location that allows the data to be mapped and analyzed. The goal of grid soil sampling is to create a map of nutrient variability and requirement for each zone in the field. The soil samples are analyzed in a laboratory for parameters such as pH, organic matter, and nutrient levels. Based on the soil test results and the crop nutrient needs, fertilizer recommendations are generated for each zone (Deonne 2021). To apply fertilizer according to the map, a variable-rate fertilizer spreader is used. This is a machine that can adjust the amount and type of fertilizer product based on the location and the map. A computer mounted on the spreader has the application map loaded into it. The computer uses the GPS receiver and the map to instruct a product delivery controller to change the fertilizer rate as the spreader moves through the field (Institute of Agriculture and Natural Resources; CropWatch 2020).

Remote-Sensing

Commercial agriculture has already adopted remote-sensing-based PA technologies, such as Crop Circle's and Green Seeker's variable fertilizer rate application technology. Over the past 10 years, the use of unmanned aerial vehicles (UAVs) has grown significantly because of their affordability and versatility in capturing the high-resolution (cm scale) images required for PA applications. Simultaneously, the abundance of satellite data has led researchers to investigate cutting edge methods for processing and storing data, like cloud computing and machine learning. For the real-time application of remote-sensing in PA, it is imperative to investigate and design a straightforward, but dependable workflow, considering the intricacy of image processing and the volume of technical knowledge and expertise required. Remote-sensing in PA can be used to assess the health of crops by measuring the plant stress caused by factors such as moisture, nutrients, compaction, diseases, and other plant health issues and used to estimate the acreage and yield of crops by analyzing the spatial and temporal variability of crop growth and production. Remote-sensing under PA can be done using different types of sensors, such as hand-held devices, aircraft-mounted cameras, or satellite-based systems (Kaushik et al., 2021; Omia et al., 2023) as follows:

- *Aerial photography*: This involves taking photographs of the land from an airplane or a drone. Aerial photographs can show the color and texture of the crops and soil, which can indicate the crop type, condition, and stress level. Aerial photographs can also show the topography and drainage patterns of the land, which can affect the crop management and productivity
- *Near-infrared (NIR) imaging*: This involves capturing images of the land using NIR wavelengths, which are invisible to the human eye but can be detected by electronic cameras. NIR images can show the amount of chlorophyll in the plant leaves, which is related to the plant health and vigor. NIR images can also show the moisture content and temperature of the soil and crops, which can affect the crop growth and development
- *Satellite remote-sensing*: This involves using satellites such as LANDSAT or SPOT to collect data from the land using different spectral bands, such as visible, NIR, and shortwave infrared (SWIR). Satellite remote-sensing can provide large-scale and frequent coverage of the land, which can be useful for monitoring crop acreage, yield, and condition over time and space. Satellite remote-sensing can also provide data on weather and climate variables, such as rainfall, temperature, and evapotranspiration, which can affect the crop water use and stress

- *GIS in agriculture*: This involves using GIS software to store, display, analyze, and integrate different types of spatial data related to agriculture, such as soil maps, Yield Maps, remote-sensing data, crop scouting reports, and soil nutrient measurements. GIS can help farmers visualize and understand the spatial variability and relationships among different factors that influence crop production and management. GIS can also help farmers evaluate and compare different management scenarios and strategies based on spatial data
- Remote-sensing is a powerful tool for agriculture that can help farmers improve their decision making and optimize their crop performance and profitability. However, remote-sensing data need to be calibrated, validated, and interpreted with ground truth data and expert knowledge to ensure their accuracy and reliability (Ennouri and Kallel 2019)

Indian Perspective of Precision Farming

The Green Revolution improved productivity, but it also caused environmental problems due to the careless use of pesticides, irrigation, and uneven fertilization. Moreover, crop production is becoming less profitable due to the decline in input efficiency and output–input ratio. Therefore, agricultural research is looking for new technologies that can meet the current and future needs and challenges. These technologies should be profitable, economical, and environmentally friendly. They should also maintain the ecological balance by using precise and site-specific management. Furthermore, planners need accurate and timely crop production forecasting systems to enhance food security. Precision farming is an approach that can address these issues, but it may not be suitable for Indian conditions. However, adoption is not impossible. Research is needed to identify the applicability of precision farming in Indian agriculture. Precision farming was initiated on an experimental basis by the M.S. Swaminathan Research Foundation (Chennai, India), in collaboration with Israel, and it included a training program. The scope of precision farming in India is as follows:

- Precision farming can help reduce production costs, increase productivity, and use natural resources more efficiently by applying inputs according to the soil, weather, and crop needs (Kumar 2023)
- Precision farming can also improve profitability, sustainability, crop quality, environmental protection, on-farm quality of life, food safety, and rural economic development by using information and communication technologies (ICTs) for farm management (Aguilar-Rivera et al., 2019)

- Precision farming can enable site-specific application of irrigation in wheat fields in Punjab and Haryana, insecticides in cotton fields, fertilizer treatments in oil palm plantations in South India, and irrigation in coffee and tea gardens. This can save water, chemicals, and energy and reduce environmental pollution (Aguilar-Rivera et al., 2019)
- Precision farming can enhance irrigation efficiency when water resources are scarce (Spanne 2021)
- Precision farming can help farmers monitor and manage pests and diseases, nutrient deficiencies, and water stress by using forecasting and mitigation tools such as sensors, drones, GPS, GIS, and remote-sensing

Precision agriculture technologies can help improve and expand the use of natural resources in India. They can lower the production costs and increase the productivity of crops. They can also enhance the profitability, sustainability, crop quality, environmental protection, on-farm quality of life, food safety, and rural economic development of farmers. PA technologies can enable farmers to apply irrigation, insecticides, and fertilizers according to the specific needs of each zone in their fields. This can save water, chemicals, and energy and reduce environmental pollution. For example, precision irrigation can be used in wheat fields in Punjab and Haryana, insecticides can be used in cotton fields, fertilizers can be used in oil palm plantations in South India, and irrigation can be used in coffee and tea gardens. PA technologies can also help farmers cope with water scarcity by improving irrigation efficiency. Moreover, PA technologies can help farmers monitor and manage pests and diseases, nutrient deficiencies, and water stress by using forecasting and mitigation tools such as sensors, drones, GPS, GIS, and remote-sensing.

Advantages of Precision Farming

- It can increase agricultural productivity and prevent soil degradation in arable land by using precise and site-specific management practices
- It can reduce the overuse of chemicals in crop production by applying them according to the soil and crop needs and conditions
- It can use water resources more efficiently by improving irrigation scheduling and application methods
- It can use GPS to map the agricultural fields and collect data on soil properties and yield variability. This can help to identify and manage the factors that affect crop performance and quality

- It can share information on agricultural practices that can improve crop quality, yield, and production at a lower cost. This can help farmers to learn from each other and adopt BMP
- It can reduce environmental risks, such as nitrate leaching and groundwater contamination, by optimizing the use of agrochemical products and minimizing their losses
- It can divide nonuniform fields into smaller zones based on their specific characteristics and requirements. This can make the field management easier and more effective
- It can offer better resource management, which can save water, energy, and chemicals and reduce waste

Disadvantages of Precision Farming

- It can have high initial costs that may discourage farmers from adopting this farming technique
- It can require expert consultation to use the PA technology effectively and appropriately
- It can take several years to collect and analyze enough data to fully implement the system and see the benefits
- It can be a complex and difficult process, especially the data collection and analysis part

Conclusion

Precision agriculture is a method of managing farms using IT to optimize the use of agricultural inputs, such as fertilizers, herbicides, tillage, and irrigation water. By using inputs more efficiently, PA can increase crop yield and/or quality without harming the environment. However, evaluating the cost benefits of PA management is not easy. Many of the technologies used in PA are still new and evolving, which makes it hard to estimate their costs and benefits. Moreover, the economic analyses of a certain technology may become outdated as the technology changes. PA can address the economic and environmental challenges of current production agriculture. The concept of "doing the right thing in the right place at the right time" is still valid, but there are still questions about how to use the technical tools available and how to measure their effectiveness.

References

Adamchuk, V.I., Hummel, J.W., Morgan, M.T., Upadhyaya, S.K. "On-the-go soil sensors for precision agriculture." *Computers and Electronics in Agriculture* 44, 1 (2004): 71-91.

Aguilar-Rivera, N., Michel-Cuello, C., Cárdenas-González, J.F. "Green Revolution and Sustainable Development." *Encyclopedia of Sustainability in Higher Education* (2019): 833-50.

Chen, F., Kissel, D.E., West, L.T., Adkins, W., Clark, R., Rickman, D., et al. "Field Scale Mapping of Surface Soil Clay Concentration." *Precision Agriculture* 5, 1 (2004): 7-26.

Deonne. 2021. Farming More Efficiently with Grid Soil Testing and Precision Nutrient Application. https://nwdistrict.ifas.ufl.edu/phag/2021/11/12/farming-more-efficiently-with-grid-soil-testing-and-precision-nutrient-application/ [Last accessed February, 2024].

Digiteum. 2021. Precision Agriculture using IoT Technology: The Future of Precision Farming. https://www.digiteum.com/precision-agriculture-technology/ [Last accessed February, 2024].

Ennouri, K., Kallel, A. "Remote Sensing: An Advanced Technique for Crop Condition Assessment." *Mathematical Problems in Engineering* (2019): 1-8.

FAO-UNESCO. (2024). https://www.fao.org/soils-portal/data-hub/soil-maps-and-databases/harmonized-world-soil-database-v12/soil-qualities-description/en/ [Last accessed February, 2024].

Farmaha, B.S., Caughman, W., Park, D. 2020. Precision Agriculture-Based Soil Sampling Strategies. https://lgpress.clemson.edu/publication/precision-agriculture-based-soil-sampling-strategies/ [Last accessed February, 2024].

Farms.com. 2024. Yield Monitoring in Precision Agriculture. https://www.farms.com/precision-agriculture/yield-monitoring/ [Last accessed February, 2024].

Gebbers, R., Adamchuk, V.I. "Precision agriculture and food security." *Science* 327, 5967 (2010): 828-31.

Institute of Agriculture and Natural Resources; CropWatch. 2002. Yield Monitoring and Mapping. https://cropwatch.unl.edu/ssm/mapping [Last accessed February, 2024].

Institute of Agriculture and Natural Resources; CropWatch. 2020. Soil Sampling for Precision Agriculture. https://cropwatch.unl.edu/ssm/soilsampling [Last accessed February, 2024].

Kaushik, S.K., Mishra, V.N., Punia, M., Diwate, P., Sivasankar, T., Soni, A.K. "Crop Health Assessment Using Sentinel-1 SAR Time Series Data in a Part of Central India." *Remote Sensing in Earth Systems Sciences* 4, 4 (2021): 217-34.

Kumar, S. 2023. Evolution Of Sustainable Farming Practices From Traditional. https://planet.outlookindia.com/opinions/evolution-of-sustainable-farming-practices-from-traditional-news-416028 [Last accessed February, 2024].

Lu, Y., Liu, M., Li, C., Liu, X., Cao, C., Li, X., et al. "Precision Fertilization and Irrigation: Progress and Applications." *AgriEngineering* 4, 3 (2022): 626-56.

Manupriya. 2021. A new digital soil map for India could help use soil optimally. https://india.mongabay.com/2021/04/a-new-digital-soil-map-for-india-could-help-use-soil-optimally/ [Last accessed February, 2024].

Mulla, D.J. "Twenty five years of remote sensing in precision agriculture: Key advances and remaining knowledge gaps." *Biosystems Engineering* 114, 4 (2013): 358-71.

Nowak, B. "Precision Agriculture: Where do We Stand? A Review of the Adoption of Precision Agriculture Technologies on Field Crops Farms in Developed Countries." *Agricultural Research* 10, 4 (2021): 515-22.

Omia, E., Bae, H., Park, E., Kim, M.S., Baek, I., Kabenge, I., et al. "Remote Sensing in Field Crop Monitoring: A Comprehensive Review of Sensor Systems, Data Analyses and Recent Advances." *Remote Sensing* 15, 2 (2023): 354.

Pandya, S., Mistry, M., Parikh, P., Shah, K., Gaharwar, G., Kotecha, K., et al. "Precision Agriculture: Methodologies, Practices and Applications." *Lecture Notes in Networks and Systems, 203 LNNS* (2021): 163-81.

Pioneer. 2019. Yield Monitor Pocket Guide. https://www.pioneer.com/us/agronomy/yield-monitor-pocket-guide.html [Last accessed February, 2024].

Sinha, N. "Precision agriculture: An overview." *Harit Dhara* 2, 2 (2019).

Spanne, A. 2021. Green Revolution: History, Technologies, and Impact. Trehugger. https://www.treehugger.com/green-revolution-history-technologies-and-impact-5189596 [Last accessed February, 2024].

8

Site-Specific Nutrient Management

Abstract

Site-specific nutrient management (SSNM) is a dynamic method that is based on plants and is usually season-specific. It combines nutrient supply and demand depending on the needs of the crop, nutrient recovery from fertilizers, and other advantageous sources with the use of geographic information system (GIS), Global Positioning System (GPS), and variable-rate technology (VRT) remote-sensing. It is a method of supplying nutrients in accordance with the requirements of their intrinsic spatial and temporal variability. The presence of nutrients can be determined using leaf color chart (LCC), SPAD, and chlorophyll meters. Low fertilization levels in fruit crops can reduce production, and excessive fertilization might injure the plants. With the development of SSNM, managing variances in soil nutrients in the field through standardized fertilizer applications has become simple. SSNM can benefit Indian farmers in controlling stress. While there is significant geographical pH variation in India, the majority of cultivated soils have an acidic character. Remote-sensing-based nutrient stress monitoring and data fusion in a GIS can aid in site-specific fertilizer and soil amendment applications, increasing fertilizer usage effectiveness and reducing nutrient losses. Although SSNM in India is still in its early stages, there is enormous potential for adoption. High-value crops have good promise for the technology. Therefore, innovative approaches can be successful if the public, private, and growers work effectively together.

Keywords: GPS, GIS, remote-sensing, SSNM

Introduction

Site-specific nutrient management (SSNM) is the dynamic, field-specific management of nutrients in a particular cropping season to optimize the supply and demand of nutrients according to their differences in cycling through soil–plant systems (Sarkar et al., 2017). This new method of nutrient recommendations is primarily based on the initial nutrient status (INS) from the soil and the crop's need for nutrients to achieve the desired yield. The SSNM recommendations could be modified to account for the inherent regional heterogeneity based on plant and soil cum plant studies for nutrient provision based on need. It keeps track of all plant nutrition sources and recommends a well-considered combination of fertilizers, manures, agricultural leftovers, and nutrient-efficient genotypes for excellent sustainability. It reduces excessive fertilizer use and aids in balancing the demand for nutrients with the supply of nutrients from organic sources, natural resources, irrigation water, and other sources. Nutrient management and recommendations are still carried out in India based on response (Dobermann and White 1998; Sarkar et al., 2017). In the 1990s, a groundbreaking agricultural innovation known as the SSNM was developed through collaborative efforts between researchers from the International Rice Research Institute (IRRI) and their partners in Asia. This innovative approach was designed to systematically analyze the geographical variations in soil fertility and the response of crops to fertilizers in diverse agricultural fields (Dobermann et al., 2004). The SSNM method introduces a scientific framework for the precise recommendation of key fertilizer elements such as nitrogen (N), phosphorus (P), and potassium (K) tailored to specific fields and their unique characteristics (Dobermann et al., 2004). The foundation of SSNM principles draws inspiration from the Quantitative Evaluation of the Fertility of Tropical Soils (QUEFTS) model, initially developed based on data from on-farm trials involving maize cultivation in Kenya (Janssen et al., 1990). The core concept behind SSNM is to calculate the required nutrient inputs for a particular field by subtracting the INS (Witt and Dobermann 2002). INS represents the quantity of a specific nutrient (N, P, or K) readily available from the soil, crop residues, irrigation water, or biological nitrogen fixation during the crop's growth cycle. This calculation is instrumental in determining the precise amount of nutrients necessary to achieve a specific target yield for the crop. Over the years, the SSNM methodology has evolved and gained significant traction, particularly in Asia. Extensive research and adaptation have led to its successful application in various crops, extending beyond rice to include wheat (Johnston et al., 2009), maize (Pasuquin et al., 2014), and several other agricultural commodities across the Asian continent. Innovations, like Nutrient Expert, have further expanded the reach and impact of SSNM. This digital platform provides SSNM solutions for 23 different crops, encompassing grain crops, cash crops, vegetables, and fruits, in China. The SSNM strategy has also found implementation in Sub-Saharan Africa (SSA), primarily targeting rice cultivation and showcasing its potential to enhance agricultural productivity in diverse settings.

Components of SSNM

Site-specific nutrient management technology works efficiently due to the interaction between its three components: information, technology, and management. The information component involves knowledge of crop characteristics, field variability, soil properties, pest population, and other factors (Rodriguez 2020). The technology component involves the use of electronic tools such as GPS (Global Positioning System) to identify the location from where data will be taken. With this, the results of soil sampling tests and yield data can be converted to maps, which is done with the help of personal computers (PCs) and geographic information systems (GIS) software. Remote-sensing techniques can also be used to detect soil variables, pest incidence, and water stress in the soil (Rodriguez 2020). The initial idea of site-specific farming is not only to measure field variability, but also to be able to apply inputs at different rates almost instantaneously, "real-time", according to the needs (Rodriguez 2020). The management component is a combination of the information we get about the field and the modern strategies available, done by the experienced farmer having a great level of knowledge about these technologies. A farmer must be capable of managing his field from the yield map, distribution weed map, moisture availability map, pest evoke map available by developing a map using GIS technology (Rodriguez 2020).

Methodology of SSNM

Site-specific nutrient management is a valuable approach for improving nutrient use efficiency, profitability, and yield in agriculture. The three foundation steps in SSNM are assessing soil and crop variability, managing the variability, and evaluating the results (Gorai et al., 2021). The available technologies help us to know the variability, by providing site-specific agronomic recommendations and manages the variability that makes precision farming valuable. Instant evaluation must be an important part of any precision farming system. Assessing variability is the critical first step in precision farming. Numbering the variability of the factors and processes and knowing when and where different setups are responsible for the spatial and temporal variation in crop yield are the challenges for precision farming. Techniques for assessing spatial variability are now available and have been applied in precision farming. The largest part of precision farming lies in assessing the spatial variability. Different technologies for assessing temporal variability also exist, but the simultaneous reporting of spatial and temporal variation is rare. We require both the spatial and temporal statistics (Dass et al., 2014). Once variation is adequately assessed, farmers must match agronomic inputs to know conditions employing management recommendations. Those are site-specific and accurate-use applications control equipment. We can adopt the technology which is most effective in site-specific variability management. While taking the soil/plant samples, we have to note the sample site coordinates, and further, we can use the same for management. This results in efficient input use and avoids excess wastage (Gorai et al., 2021). There are three essential problems regarding precision farming evaluation: economics, environment, and technology transfer. The most initial fact regarding the analysis of profits of precision farming is that the value comes from the application of the data and not technology. Assessment of soil and crop variability, managing the variability, and its evaluation are three foundation steps in SSNM. The available technologies help us in knowing the variability and by providing site-specific agronomic recommendations, we can manage the variability that makes precision farming valuable and instant evaluation must be an important part of any precision farming system.

Tools Used in SSNM

Optical Sensors

Farmers and experts can use optical sensors to develop SSNM recommendations, particularly for nitrogen (Ahmad and Mahdi 2018). Optical sensors are used to measure reflectance from the leaves to generate a vegetative index called normalized difference vegetation index (NDVI), which measures the nutrient status of the plants based on their size and color (green versus yellow). The application of optical sensor-based nutrient management requires a local calibration from the sensor for a given nutrient, crop, and region. This calibration helps in providing the grain yield of the crop to the NDVI readings (Ahmad and Mahdi 2018). Once this is completed, optical sensors require:

- Settlement of a reference strip in the farmer's field that will receive a considerable amount of nitrogen.
- Collection of an NDVI reading in the reference strip helps the farmer needs to know how much nitrogen should be applied.
- The NDVI readings collected from these two areas in the field at the same time with the date of sowing and date of sensing are entered in a mathematical model developed for every region.

Software-based Tools

Computer or mobile phone-based tools are increasingly used to improve nutrient management practices in farmers' fields, especially in geographies where blanket fertilizer recommendations are there (Ahmad and Mahdi 2018). These tools provide small-scale maize, rice, and wheat farmers with crop and nutrient management advice customized to their farming conditions and requirement (Ahmad and Mahdi 2018). One such tool is Nutrient Expert, which is a computer-based decision-support tool that helps smallholder farmers to rapidly implement SSNM in their individual fields with or without soil test data. The software measures the attainable yield for a farmer's field based on the growing conditions, estimating the nutrient balance in the cropping system based on yield and fertilizer/manure applied in the previous crop and combines such information with expected nitrogen, phosphorus, and potassium response in target fields to generate location-specific nutrient recommendations (Chivenge et al., 2021). Another tool is Crop Manager, which is a computer- and mobile phone-based application that helps small-scale rice, rice–wheat, and maize farmers with site- and season-specific recommendations for fertilizer application. The tool helps farmers to adjust nutrient application to crop needs based on soil characteristics, water management, and crop variety on their farm. Recommendations are based on user input information about farm location and management, which can be collected by extension workers, crop advisors, and service providers (Kumari 2018). Nutrient Expert is a computer-based decision-support tool that helps smallholder farmers to rapidly implement SSNM in their individual fields with or without soil test data (Kumari 2018). The software measures the attainable yield for a farmer's field based on the growing conditions, estimating the nutrient balance in the cropping system based on yield and fertilizer/manure applied in the previous crop and combines such information with expected nitrogen, phosphorus, and potassium response in target fields to generate location-specific nutrient recommendations (Kumari 2018).

Soil Plant Analysis Development

A SPAD threshold value of 35 has been found to work well in transplanted rice, while a higher threshold value near 40 is required for semidwarf wheat varieties (Varinderpal-Singh et al., 2010). The SPAD value is measured four to five times in a season at 10-day intervals starting from the tillering stage (Varinderpal-Singh et al., 2010).The use of SPAD meters has been shown to improve nitrogen management in crops such as rice and wheat, leading to increased yields and reduced fertilizer use (Balasubramanian et al., 1998). The chlorophyll meter (SPAD meter) is a simple portable diagnostic tool used to monitor crop nitrogen status in the field. When calculated to local important crop varieties and growing conditions of the crop, it serves as an efficient tool for developing need-based variable-rate nitrogen application for cereal crops (Balasubramanian et al., 1998). The method involves measuring dimensionless "SPAD value", which is then compared with a critical threshold value to decide whether and how much nitrogen needs to be applied (Balasubramanian et al., 1998).

Leaf Color Chart

The leaf color chart (LCC) is another simple, easy-to-use, and inexpensive tool to maximize nitrogen use efficiency in rice and other cereal crops (Bhat et al., 2022). The chart contains six green strips with the color ranges from yellow–green to dark green (Rao and Das 2023). It has been calibrated with the chlorophyll meter, which is used to help in nitrogen top-dressing in rice (Varinderpal-Singh et al., 2010). The critical leaf color reading for nitrogen top-dressing ranges from three for varieties with light green foliage of scented rice varieties to four for semidwarf *indica* varieties and five for hybrid rice varieties (Bhat et al., 2022; Rao and Das 2023). To summarize, the leaf color chart is a reliable tool for nitrogen management in rice and other cereal crops. It contains six green strips with the color ranges from yellow-green to dark green and has been calibrated with the chlorophyll meter. The critical leaf color reading for nitrogen top-dressing ranges from three for varieties with light green foliage of scented rice varieties to four for semidwarf *indica* varieties and five for hybrid rice varieties.

Applications of SSNM

- This provides a regional and locale specific nutrient-based management tailored to the field and season-specific needs for a crop Site specific nutrient management (SSNM) recommendations could be modified to account for the inherent regional heterogeneity based on plant and soil cum plant studies for nutrient provision based on need (Ahmad and Mahdi 2018; Dobermann and White 1998; Sarkar et al., 2017). Efficient nutrient management is crucial for the sustainable production of rice in Asia. The use of already-existing native nutrient sources, such as crop leftovers, is an important strategy for nutrient management in irrigated and rainfed lowland rice systems in Asia (Ayilara et al., 2020). The supply of N, P, and K fertilizers is adapted to the necessary areas and in accordance with the unique requirements of the crop (Ayilara et al., 2020). Using a leaf control chart (LCC) ensures that nitrogen is used as effectively as possible and thereby minimizes nitrogen waste (Ayilara et al., 2020). Local randomization is used for applying Zn, S, and micronutrients (Ayilara et al., 2020). Optimizing the soil nutrient delivery over time and space to meet crop needs is another important strategy (Ayilara et al., 2020). This can be achieved through SSNM strategies, which are tailored to the large between-field differences in indigenous nutrient supply. The iterative procedure proposed includes: (1) estimation of the potential indigenous supplies of N (INS), P (Information Production System), and K (IKS, all in kg ha^{-1}) and diagnosis of other nutritional disorders in year one; (2) estimation of a field- or farm-specific recommendation for fertilizer use based on nutrient interactions and economic yield target; (3) optimization of timing and amount of applied N based on actual plant growth; (4) estimation of the change in INS, IPS and IKS based on the nutrient balance after harvest; and (5) use of the adjusted INS, IPS, and IKS as model inputs for the subsequent rice crop

- *For yield increment*: SSNM keeps track of all plant nutrition sources and recommends a well-considered combination of fertilizers, manures, agricultural leftovers, and nutrient-efficient genotypes for excellent sustainability. It reduces excessive fertilizer use and aids in balancing the demand for nutrients with the supply of nutrients from organic sources, natural resources, irrigation water, and other sources (Ahmad and Mahdi 2018; Dobermann and White 1998; Sarkar et al., 2017)
- *To hike fertilizer use efficiency*: SSNM reduces excessive fertilizer use and aids in balancing the demand for nutrients with the supply of nutrients from organic sources, natural resources, irrigation water, and other sources (Ahmad and Mahdi 2018; Dobermann and White 1998; Sarkar et al., 2017)
- *To increase profitability*: SSNM recommends a well-considered combination of fertilizers, manures, agricultural leftovers, and nutrient-efficient genotypes for excellent sustainability. It reduces excessive fertilizer use and aids in balancing the demand for nutrients with the supply of nutrients from organic sources, natural resources, irrigation water, and other sources (Ahmad and Mahdi 2018; Dobermann and White, 1998; Sarkar et al., 2017)
- *To improve marketable crop quality*: Low fertilization levels in fruit crops can reduce production, and excessive fertilization might injure the plants. With the development of SSNM, managing variances in soil nutrients in the field through standardized fertilizer applications has become simple (Ahmad and Mahdi 2018; Dobermann and White 1998; Sarkar et al., 2017)

To combat environmental stress: Reducing greenhouse gas emissions from agriculture in regions where nitrogen fertilizer is widely used is a key goal (Ayilara et al., 2020). This can be achieved by reducing the use of nitrogen fertilizers and adopting alternative practices such as the use of biofertilizers, which are derived from organic matter and are environmentally friendly (Itelima et al., 2018). Reducing the buildup of inorganic nutrients when the crop is in low demand is also an important consideration . This can be achieved by using slow-release fertilizers, which release nutrients over a longer period of time, or by using cover crops, which absorb excess nutrients from the soil (Itelima et al., 2018).

- Remote-sensing-based nutrient stress monitoring and data fusion in a GIS can aid in site-specific fertilizer and soil amendment applications, increasing fertilizer usage effectiveness and reducing nutrient losses. SSNM can benefit Indian farmers in under stress (Dobermann and White 1998; Sarkar et al., 2017)

SSNM and Crop Response

The adoption of SSNM has brought about significant agricultural benefits in the Indian subcontinent, particularly in the cultivation of rice, maize, and cassava in SSA (Jat et al., 2013; Ortiz-Monasterio and Raun 2007; Pasuquin et al., 2014; Peng et al., 2010; Singh et al., 2015). This innovative approach has proven to enhance nutrient utilization efficiency, overall profitability, and crop yields, thereby contributing to the sustainable improvement of agricultural practices in the region. Research findings indicate the substantial advantages of SSNM over conventional farming methods. In the case of rice and maize, SSNM has demonstrated remarkable yield increases, with gains of 24% and 69%, respectively, outperforming conventional practices (Jat et al., 2013; Ortiz-Monasterio and Raun 2007; Pasuquin et al., 2014; Peng et al., 2010; Singh et al., 2015). Moreover, when compared to local guidelines, SSNM has consistently led to higher yields, with maize and rice production achieving 11% and 4% higher yields, respectively. The adaptability and scalability of SSNM tools have become increasingly evident through collaboration with digital platforms and databases, enhancing their utility and precision. Notably, studies conducted across Southeast Asian locations revealed that SSNM can boost grain yields by 13% over a 3-year period, even though yields may initially marginally decline. Furthermore, a comprehensive review of 179 rice farms in six Asian countries confirmed that SSNM consistently results in increased yields. In South Asia, SSNM has also shown its potential in wheat systems, contributing to significant grain yield improvements ranging from 18% to 27% compared to traditional fertilizer practices. Additionally, research in Chinese rice fields demonstrated a 5% increase in grain yields under SSNM, attributed to optimized nitrogen inputs, leading to reduced insect damage and disease (Jat et al., 2013; Ortiz-Monasterio and Raun 2007; Pasuquin et al., 2014; Peng et al., 2010; Singh et al., 2015). One of the economic advantages of SSNM lies in its ability to improve overall profits for farming enterprises. By optimizing fertilizer use, SSNM can save farmers money on fertilizer expenses, although the extent of these savings depends on various factors, including baseline yields, initial fertilizer usage, and fertilizer costs. For instance, experiments with optical sensors in wheat and barley fields in Mexico showed a savings of 40–70 kg of nitrogen per hectare without compromising grain yield. Similarly, wheat production experiments in India indicated a substantial increase in net returns from USD 390 to USD 1,071 per hectare, although labor costs increased by USD 123 per hectare. These findings underscore the potential of SSNM to revolutionize agricultural practices in the Indian subcontinent, promoting sustainable, efficient, and profitable farming while also ensuring food security and environmental stewardship (Jat et al., 2013; Ortiz-Monasterio and Raun 2007; Pasuquin et al., 2014; Peng et al., 2010; Singh et al., 2015).

A site-specific approach to nutrient management examined in 56 on-farm experiments with irrigated wheat and transplanted rice crops in North-West India revealed that field-specific management of various macronutrients maximized the yields of rice and wheat crops by 12% and 17%, respectively, and increased profitability by 13–14% in the North-West region of India (Verma et al., 2020). In maize, SSNM maximizes the agronomic efficiency of nitrogen fertilizer by 53% compared to FFP. Abdulrachman *et al.* reported that the results of 45 trials conducted on irrigated rice in different countries of South Asia reveal that with SSNM, fertilizer nitrogen rates were reduced significantly, i.e., by 10–20% (Abdulrachman et al., 2002).Chlorophyll meter-based nitrogen application (30 kg basal + 30 kg N/ha at SPAD value 37.5) saved 30 kg N/ha and increased rainy season maize grain yield by 10% when compared to soil test-based nitrogen application in the semiarid north plain zone of India (Ahmad and Mahdi 2018; Verma et al., 2020). Thus, a site-specific approach to nutrient management maximizes the yields of rice and wheat crops by 12% and 17%, respectively, and increases profitability by 13–14% in the North-West region of India. SSNM maximizes the agronomic efficiency of nitrogen fertilizer by 53% compared to FFP in maize. Chlorophyll meter-based nitrogen application saved 30 kg N/ha and increased rainy season maize grain yield by 10% when compared to soil test-based nitrogen application in the semiarid north plain zone of India (Table 1).

Challenges for SSNM

- SSNM requires knowledge of different soil properties and the ability to monitor crops' nutrient status and adjust fertilizer inputs accordingly (Verma et al., 2020). While the need to conduct on-farm nutrient trials and soil tests has historically been a problem for its implementation, the use of decision support systems and farmer-friendly tools and techniques that use similar information to calculate nutrient requirements make SSNM more accessible to farmers (Verma et al., 2020)

- The availability of fertilizers, whether synthetic or organic, is not universal. The establishment of input markets or identification of on-farm nutrient sources may be necessary (Rodriguez 2020)
- For SSNM to increase farmers' profits, it must deliver either (1) savings from reduced fertilizer use without a decrease in yields, or (2) yield rise that are valued higher than the costs of acquirement by using SSNM technology. Farmers are more likely to see positive net returns with high-value crops where yield increases can substantially increase profits, or when fertilizer prices are high (Ahmad and Mahdi 2018). Hence, SSNM requires knowledge of different soil properties and the ability to monitor crops' nutrient status and adjust fertilizer inputs accordingly. The use of decision support systems and farmer-friendly tools and techniques that use similar information to calculate nutrient requirements make SSNM more accessible to farmers. The availability of fertilizers, whether synthetic or organic, is not universal

Conclusion

Site-specific nutrient management is a dynamic and innovative approach to optimizing nutrient management in agriculture. This method, based on plants and tailored to specific seasons, combines technology, information, and management to provide region-specific nutrient recommendations.

Table 1: Effect of site-specific nutrient management (SSNM) on wheat productivity (t ha^{-1}) and economic returns (Rs ha^{-1}, in parenthesis) at seven locations in India.

Site	FP	SR	SSNM	Increase over SR (% [Rs ha^{-1}])	Increase over FP (% [Rs ha^{-1}])
Ranchi	2.56 (1,575)	4.15 (25,276)	4.06 (26,854)	10.0 (1,578)	58.5 (25,309)
Modipuram	4.77 (29,292)	4.90 (31,859)	6.43 (58,083)	31.0 (26,224)	46.5 (28,791)
Kanpur	4.72 (7,258)	5.45 (17,644)	6.00 (31,338)	10.1 (13,694)	27.1 (24,080)
Ludhiana	5.45 (27,772)	6.28 (39,105)	6.55 (46,219)	4.3 (7,114)	20.1 (18,447)
Sabour	3.92 (18,306)	4.97 (28,614)	5.82 (45,116)	17.1 (16,502)	48.7 (26,810)
Pantnagar	3.87 (7,828)	5.10 (14,276)	6.39 (19,426)	25.3 (5,150)	66.0 (11,598)
Palampur	2.64 (55,122)	3.76 (54,583)	3.87 (60,905)	3.0 (6,322)	46.5 (5,783)

(*FP*: farmers' practice; SR: state fertilizer recommendation)

Source: Singh and Bansal 2010

SSNM offers several significant benefits, including the ability to enhance yield, increase profitability, improve crop quality, and combat environmental stress. It helps reduce excessive fertilizer usage and optimize nutrient supply according to the unique needs of each field. SSNM has evolved over the years and has shown promising results in various crops, particularly in Asian and SSA regions. The adoption of SSNM has led to significant yield increases and cost savings for farmers, making it a valuable tool for sustainable and efficient agriculture. Various tools, such as optical sensors, software-based solutions like Nutrient Expert and Crop Manager, SPAD meters, and LCCs, have been instrumental in implementing SSNM. It represents a crucial step forward in precision agriculture, offering farmers the opportunity to optimize nutrient management, increase crop yields, and improve overall profitability while contributing to sustainable and environmentally friendly farming practices. As this innovative approach continues to evolve and adapt to different regions and crops, it holds great promise for enhancing global food security and agricultural sustainability. Collaboration between the public, private sector, and growers will be essential in realizing the full potential of SSNM

References

Abdulrachman, S., Gines, H.C., Nagarajan, R., Satawathananont, S., Son, T.T., Tan, P.S., et al. "Variation in the Performance of Site-Specific Nutrient Management among Different Environments with Irrigated Rice in Asia." *Better Crops International* 16, 2 (2002): 18.

Ahmad, L., Mahdi, S.S. 2018. "Site-Specific Nutrient Management." *Satellite Farming: An Information and Technology Based Agriculture.* Germany: Springer International Publishing. pp. 97-109.

Kumari, P. 2018. Precision nutrient management tools help Indian farmers make smart fertilizer decisions. https://ricetoday.irri.org/precision-nutrient-management-tools-help-indian-farmers-make-smart-fertilizer-decisions/ [Last accessed February, 2024].

Ayilara, M.S., Olanrewaju, O.S., Babalola, O.O., Odeyemi, O. "Waste Management through Composting: Challenges and Potentials." *Sustainability* 12, 11 (2020): 4456.

Balasubramanian, V., Morales, A.C., Cruz, R.T., Abdulrachman, S. "On-farm adaptation of knowledge-intensive nitrogen management technologies for rice systems." *Nutrient Cycling in Agroecosystems* 53, 1 (1998): 59-69.

Bhat, T.A., Kanth, R.H., Jan, B., Nazir, A., Ahanger, S.A., Mir, M.S., et al. Real-Time Nitrogen Application of Rice Varieties Based on Leaf Colour Chart under System of Rice Intensification in Temperate Climate. *Agronomy* 12, 9 (2022): 2229.

Chivenge, P., Sharma, S., Bunquin, M.A., Hellin, J. "Improving Nitrogen Use Efficiency—A Key for Sustainable Rice Production Systems." *Frontiers in Sustainable Food Systems* 5 (2021).

Dass, A., Suri, V.K., Choudhary, A.K. "Site-specific nutrient management approaches for enhanced nutrient-use efficiency in agricultural crops." *Research and Reviews: Journal of Crop Science and Technology* 3, 3 (2014): 1-6.

Dobermann, A., White, P.F. "Strategies for nutrient management in irrigated and rainfed lowland rice systems." *Nutrient Cycling in Agroecosystems* 53, 1 (1998): 1-18.

Dobermann, A., Witt, C., Dawe, D. 2004. *Increasing Productivity of Intensive Rice Systems through Site-specific Nutrient Management.* Philippines: IRRI; Science Publishers, Inc.

Gorai, T., Yadav, P.K., Choudhary, G.L., Kumar, A. "Site-specific Crop Nutrient Management for Precision Agriculture—A Review." *Current Journal of Applied Science and Technology* 40, 10 (2021): 37-52.

Itelima, J.U., Bang, W.J., Onyimba, I.A., Sila, M.D., Egbere, O.J. "Bio-fertilizers as Key Player in Enhancing Soil Fertility and Crop Productivity: A Review." *Direct Research Journal of Agriculture and Food* Science 6, (3): 73-83.

Janssen, B.H., Guiking, F.C.T., van der Eijk, D., Smaling, E.M.A., Wolf, J., van Reuler, H. "A system for quantitative evaluation of the fertility of tropical soils (QUEFTS)". *Geoderma* 46, 4(1990): 299-318.

Jat, M.L., Satyanarayana, T., Majumdar, K., Parihar, C.M., Jat, S.L., Tetarwal, J.P., et al. "Fertiliser Best Management Practices for Maize Systems." *Indian Journal of Fertilisers* 9, 4 (2013): 80-94.

Johnston, A.M., Khurana, H.S., Majumdar, K., Satyanarayana, T. "Site-specific nutrient management–concept, current research and future challenges in Indian agriculture." *Journal of the Indian Society of Soil Science* 57, 1 (2009): 1-10.

Ortiz-Monasterio, J.I., Raun, W. "Reduced nitrogen and improved farm income for irrigated spring wheat in the Yaqui Valley, Mexico, using sensor based nitrogen manageme." *The Journal of Agricultural Science* 145, 3 (2007): 215-22.

Pasuquin, J.M., Pampolino, M.F., Witt, C., Dobermann, A., Oberthür, T., Fisher, M.J., et al. "Closing yield gaps in maize production in Southeast Asia through site-specific nutrient management." *Field Crops Research* 156 (2014): 219-30.

Peng, S., Buresh, R.J., Huang, J., Zhong, X., Zou, Y., Yang, J., et al. "Improving nitrogen fertilization in rice by sitespecific N management. A review." *Agronomy for Sustainable Development* 30, 3 (2010): 649-56.

Rao, K.G., Das, N.K. "Leaf color chart (LCC) a reliable tool for nitrogen management in (Dry seeded rice and transplanted rice): A review." *The Pharma Innovation Journal* 12, 6 (2023): 3974-9.

Rodriguez, D.G.P. "An Assessment of the Site-Specific Nutrient Management (SSNM) Strategy for Irrigated Rice in Asia." *Agriculture* 10, 11 (2020): 559.

Sarkar, D., Meena, V.S., Haldar, A., Rakshit, A. "Site-Specific Nutrient Management (SSNM): A Unique Approach Towards Maintaining Soil Health." *Adaptive Soil Management: From Theory to Practices* Singapore: Springer. pp. 69-88.

Singh, V.K., Shukla, A.K., Dwivedi, B.S., Singh, M.P., Majumdar, K., Kumar, V., et al. "Site-Specific Nutrient Management under Rice-Based Cropping Systems in Indo-Gangetic Plains: Yield, Profit and Apparent Nutrient Balance." *Agricultural Research* 4, 4 (2015): 365-77.

Varinderpal-Singh, Bijay-Singh, Yadvinder-Singh, Thind, H.S., Gupta, R.K. "Need based nitrogen management using the chlorophyll meter and leaf colour chart in rice and wheat in South Asia: a review." *Nutrient Cycling in Agroecosystems* 88, 3 (2010): 361-80.

Verma, P., Chauhan, A., Ladon, T. "Site-Specific Nutrient Management." *Journal of Pharmacognosy and Phytochemistry* 9, 5 (2020): 233-6.

Witt, C., Dobermann, A. "A site-specific nutrient management approach for irrigated, lowland rice in Asia." *Better Crops International* 16, 1 (2002): 20-4.

9

Fertilizer Recommendation Based on Variable Rate Technology

Abstract

Three key elements contribute to successful crop growth: Climate, soil, and water. For optimal financial returns and efficient use of resources, these should be economically managed alongside other inputs like fertilizers and insecticides. Over the years, agricultural management techniques and technology have seen steady progress. Yet, in Asian countries, there is still a reliance on low-efficiency practices, indicating a potential for technological innovation. Variable-rate technology (VRT) could be particularly advantageous for developing nations such as India. However, the implementation of this technology requires farmers' fields to be receptive to specific interventions and land consolidation procedures.

Keywords: Efficiency, potency, VRT, consolidation

Introduction

To address the escalating food demands of the global population, a transformation in the agricultural sector is imperative (Velten et al., 2015; Lampridi et al., 2019). The current overdependence on finite resources is leading to their overexploitation, largely due to irresponsible usage (Velten et al., 2015; Lampridi et al., 2019). By adopting a site-specific approach, farmers can apply inputs only where they are necessary in the field (Velten et al., 2015; Lampridi et al., 2019). This method can enhance output, decrease labor costs, and mitigate the environmental damage caused by excessive input use (Velten et al., 2015; Lampridi et al., 2019). Variable-rate technology (VRT) is a tool that allows farmers to apply materials such as fertilizer, water, chemicals, and seed at different rates within a field (He 2022). With VRT, we can adjust the amount of fertilizer applied based on the initial soil nutrient content of each distinct plot of land (He 2022). Consequently, certain areas of the same field receive less or no fertilizer at all, while other areas receive more. When it comes to choosing between uniform rate technology (URT) and VRT for crop production on land with the aforementioned characteristics, the decision ultimately hinges on the economic costs and benefits (He 2022).

Farmers typically adhere to recommendations derived from studies and field experiments conducted under specific agroclimatic conditions, which are then extrapolated to a regional scale (Mistary 2021). These recommendations guide the use of various agricultural inputs, such as seeds, fertilizer, herbicides, pesticides, and water (Mistary 2021). However, when cultivating a particular crop, it is crucial to consider the inherent variability in factors like soil types, soil nutrition, weed density, pest infestations, and soil water. This variability exists not only between different locations and farms but also between and within individual fields or plots (Mistary 2021). VRT can enhance the efficiency of applied inputs, minimizing misuse that could lead to environmental degradation (He 2022). Tools like the Geographic Information System (GIS) and Global Positioning System (GPS) have been employed to manage preexisting field variability, leading to the development of VRT (Tayari et al., 2015). VRT can increase input efficiency by reducing input misuse, thereby limiting the risk of pesticide and fertilizer leaching into water sources and reducing application in environmentally sensitive locations (He 2022). As a result, it aids in boosting agricultural output by optimizing resource utilization (He 2022). VRT allows for the application of fertilizer, chemicals, lime, gypsum, irrigation water, and other farm inputs at varied rates across a field, eliminating the need for manual adjustments to equipment rate settings. By responding to fluctuations in crop yield potential with precise input rates, VRT increases system efficiency and reduces the likelihood of environmental issues (He 2022).

VRT Rationale

Variable-rate technology continues to play pivotal role in contemporary agriculture (He 2022). With the surge in population growth, there has been a corresponding increase in the demand for agricultural products, such as crop cereals and livestock feed (Maja and Ayano 2021; Ritchie et al., 2018). This has led to an overexploitation of land resources, resulting in soil degradation (AbdelRahman 2023; Bhattacharyya et al., 2015; Hossini et al., 2022; Teague and Kreuter 2020). Consequently, a multitude of strategies have been and continue to be employed to optimize soil use for agriculture and other sectors with VRT being one of these techniques (He 2022). When cultivating certain crops, it is crucial to consider the inherent variability in factors like soil types soil nutrition, weed density, pest infestations, and soil water. This variability exists not only between different locations and farms, but also between and within individual fields or plots. The application of inputs can be made more efficient using VRT, thereby minimizing misuse that could lead to environmental deterioration (He 2022). North America is currently at the forefront of VRT due to the region's significant advancements in agricultural automation and digitalization (MacPherson et al., 2022). However, the Asia Pacific region, with its vast farmland and population, is expected to see substantial growth in this area between 2021 and 2028 (Jun et al., 2022). Developing countries such as India, China, Brazil, Argentina, and Malaysia have begun to implement precision agriculture practices, particularly the VRT technique, primarily on research farms (Mohapatra 2022; Rahate and Khandekar 2018). Despite this, the adoption of these practices is still in its nascent stages. Therefore, it is crucial to closely monitor the adoption trend and focus research efforts in this area (Mohapatra 2022; Rahate and Khandekar 2018).

Types of VRT

Variable Rate Technology can be effectively utilized to address the spatial variability that exists between different fields or within distinct management zones/classes (He 2022). VRT is primarily categorized into two types:

- *Map-based VRT*: Also known as prescription map-based VRT, this method involves altering the application rate based on an electronic map (He 2022). The field position is determined using a GPS receiver, which aids in constructing a specific map of the field. This map is then integrated into the system and modified as the applicator traverses through the field
- *Sensor-based VRT*: This type does not require a location system or a map (He 2022). The applicators' sensors monitor the properties of the crop or the soil. A control system then assesses the input needs of the soil or plants based on the collected data, and an input is supplied to the location where the sensor was used for measurement
- *Tools and techniques used in VRT*: Precision agriculture focuses on automatically applying material to a specific area (He 2022). Data gathered by sensors, maps, and GPS is used to determine how the substance is applied. These components, which help maximize agricultural output, include items like fertilizer, chemicals, and seeds. Precision agriculture uses a wide range of technological tools. They range from drones and satellites to artificial intelligence and hyperspectral photography, among other technologies. Whichever variable-rate application technique

is employed, it is critical to comprehend the fundamental principles involved

- *Sensors*: There are several different types of sensors, including GPS and DGPS, are two examples of positioning systems, the most widely used variety (He 2022). With the best implementations, it helps to enhance location precision from nominal GPS accuracy of 15 m to roughly 10 cm (He 2022)
    - Soil and plant sensors use the organic content of the soil and light refraction from the soil to calculate soil health (He 2022). It aids in the evaluation of electrical resistance and light reflectance, enabling simple moisture estimation (He 2022)
    - *Pressure sensors*: Change the spray pattern by producing an electronic signal corresponding to the fluid pressure used in the fluid applicator structure (He 2022)
    - Flow sensors assess how much fluid passes through a pipe or hose in a given amount of time (He 2022). They can monitor the flow of clean liquids or slurries (He 2022). Some people use gallons per minute to estimate volumetric flow rate (He 2022). Sensor for speed tachometer or sensors for shaft speed. Measure the shaft's rotational speed; this is useful for managing shaft speed but not for determining the true speed of the ground (He 2022)
- *Controllers*: Controllers play a crucial role in reading sensor inputs and calculating product output, which largely depends on factors such as soil texture, soil nutrients, yield, etc (He 2022). They also assist in adjusting the rate at which products are applied (He 2022)
- *Actuators*: Actuators respond to the signals from controllers (He 2022). They control the quantity of product applied to fields (He 2022). They respond to electrical, pneumatic, and hydraulic signals. Their functions include extending or retracting, opening or closing a gate, adjusting speed, or changing the position of a valve (He 2022)
- *Data imagery used*: After deciding whether to employ a map-based or sensor-based strategy, the next step is to decide what kind of data the sensor should be collecting or what kind of images should be used in the mapping (He 2022). In order to obtain information about the environment, VRA technology usually uses drones or other imaging devices (He 2022). Additional sensors include N sensors from Yara, Isaria from Fritzmeier, and Green Seeker for Trimble. Information that

is essential for fertilizer application includes details about the soil, the type of crop being grown, climate data, and the vehicle's speed. All of this information is made available through the usage of the variable-rate application technology

Crop Performance Under VRT

Sugarcane is cultivated on approximately 20.3 million acres of land globally, with Brazil accounting for 24% of this total (Rossetto et al., 2022). The Brazilian sugarcane industry, which places a strong emphasis on adjusting tillage depth according to soil texture to minimize the need for expensive direct soil inputs, is well-positioned to adopt VRT and soil erosion management strategies (Rossetto et al., 2022). In Brazil, Magahales and Cerri developed VRT and other precision agriculture technologies (Rossetto et al., 2022). The system proved to be reliable and consistent, as evidenced by the data, which showed a correlation of 0.66 between the outputs of the sugarcane yield monitor and the weight of the harvested load at the time (Rossetto et al., 2022). The fact that site-specific Zn application trial results in Uruguay demonstrated that the profit could reach up to $50 per acre highlights the benefits of this approach (Liu et al., 2019). Furthermore, a study using precision agriculture on banana farms was conducted in Costa Rica (Zamykal and Everingham 2009). This technology allows farmers to connect to a soil database through yield monitoring and make decisions about soil fertility and disease-related issues relevant to their soil (Zamykal and Everingham 2009). India currently produces 200 Mt of food grain, making the country self-sufficient in food (Halawar 2019). However, meeting the quantity demand alone cannot satisfy the needs of global agricultural markets (Halawar 2019). Out of the 115.6 million farms, 107.08 million have a land size of <4 hectares (Halawar 2019). Therefore, the government is required to take some action to organize dynamic soil sampling in order to create nutrient maps using already available technology for this vast number of farms (Padbhushan et al., 2022). By predicting intermediate conditions, the accurately validated zone-specific computer simulation model can extend the time between two subsequent soil samplings (Ahmed et al., 2022). The panchayats (village regulatory body) can distribute these nutrient maps along with easy-to-understand fertilizer recommendations for each management zone within the field (Ahmed et al., 2022).

VRT Under Indian Context

Indian agriculture, predominantly led by small and marginal farmers, faces numerous constraints. These farmers often cannot afford capital inputs such as machinery and tractors or adopt new technology for crop production

(Randive et al., 2021). As a result, they resort to using consumable modern inputs like agrochemicals in the form of fertilizers and pesticides to enhance their productivity (Randive et al., 2021). Subsidies in this sector are further exacerbating its use at 4.16 per cent per annum, besides imbalancing the nutrient ratio of N:P:K to 6.7:2.7:1 from the ideal requirement of 4:2:1 (Randive et al., 2021). India ranks second in the world and first among the SAARC (South Asian Association for Regional Cooperation) countries in terms of total fertilizer consumption, with an average per hectare application of about 145 kg (Randive et al., 2021). This has resulted in reduced nitrogen use efficiency (NUE) of 15–30%, and low phosphorus use recovery of 15% (Randive et al., 2021). It is projected that the rate of consumption may increase to 277 kg/ha (Randive et al., 2021). The leading states of fertilizer consumption in India are Telangana (245.3 kg/ha), Haryana (224.5 kg/ha), and Punjab (224.5 kg/ha) (Randive et al., 2021). In contrast, Jammu and Kashmir has a consumption of 61.9 kg/ha (CEIC 2021). However, the scenario of fertilizer use in Jammu

Table 1: All India consumption of N, P_2O_5, and K_2O (1950–51 to 2020–21).

ALL India consumption of N, P_2O_5, and K_2O (1950–51 to 2020–21) (000 tones)				
Year	**N**	**P_2O_5**	**K_2O**	**Total (N + P_2O_5 + K_2O)**
1950–51	55.0	8.8	6.0	69.8
1960–61	211.7	53.1	29.0	293.8
1970–71	1,479.3	541.0	236.3	2,256.6
1980–81	3,678.1	1,213.6	623.9	5,515.6
1990–91	7,997.2	3,221.0	1,328.0	12,546.2
2000–01	10,920.2	4,214.6	1,567.5	16,702.3
2010–11	16,558.2	8,049.7	3,514.3	28,122.2
2019–20	19,101.3	7,662.0	2,607.0	29,370.4
2020–21 (P)	20,404.0	8,977.9	3,153.7	32,535.6

***Inference*: All India Consumption of N, P_2O_5 and K_2O (1950–51 to 2020–21) (`000 tones)**				
	N	**P**	**K**	**Total**
Pre-Green Revolution Period (1950–61)	133.35	35.35	17.5	186.2
Post-Green Revolution period (1970–91)	4,384.8	2,828.2	729.4	7,942.4
Year 2000–2020	15,533.13	6,642.10	2,562	24,737.23

Table 2: State-wise (kg/ha) consumption of fertilizer (N + P + K).

State/union territory/zone	2018–19
Andhra Pradesh	173.3
Telangana	245.3
Karnataka	183.2
Kerala	36.4
Tamil Nadu	186.4
Puducherry	227.9
Andaman and Nicobar Islands	16.1
Lakshadweep	–
Gujarat	135.5
Madhya Pradesh	90.3
Chhattisgarh	86.3
Maharashtra	126.0
Rajasthan	60.8
Goa	36.5
Daman and Diu	38.6
Dadra and Nagar Haveli	13.9
Haryana	224.5
Punjab	224.5
Uttar Pradesh	170.1
Uttaranchal	140.7
Himachal Pradesh	63.3
Jammu and Kashmir	61.9
Bihar	227.3
Jharkhand	59.8
Odisha	70.6
West Bengal	161.1
Assam	73.7
Tripura	0.0

(Contd...)

Table 2: *(Contd...)*

State/union territory/zone	**2018–19**
Manipur	68.3
Meghalaya	0.0
Nagaland	–
Arunachal Pradesh	–
Mizoram	55.9
Sikkim	–
ALL INDIA (Average)	133.1
Inference of highest fertilizer consuming state (kg/ha)	**2018–19**
Telangana	245.3
Haryana	224.5
Punjab	224.5

Source: Anonymous 2020, RBI report, Agricultural Statistics At a Glance, Ministry of Agriculture and Farmers Welfare, Government of India

and Kashmir is increasing at a rate of 10–15 % every year. This excessive use of chemical fertilizers leads to problems such as algae eutrophication, environmental degradation, and climate change (Randive et al., 2021). Adoption of Integrated Nutrient Management (INM) practices, use of neem-coated urea, adoption of location-specific crops and genotypes can reduce the fertilizer use by 20–30% (Juttu et al., 2021; Reddy 2016). These practices not only enhance the productivity but also contribute to the sustainability of the agricultural system (Juttu et al., 2021; Reddy 2016).

Challenges and Future Strategies for VRT in India

The implementation of VRT in agriculture faces several challenges. One of the primary obstacles is the varied farming practices of farmers and their small, dispersed land holdings (Chaudhary et al., 2020). Other difficulties that restrict the use of VRT include the high initial cost of machinery and the expense of maintaining and operating it (Chaudhary et al., 2020; He 2022). There is also a shortage of technical expertise and a weak network of suppliers for purposes of tailored hiring (Chaudhary et al., 2020; He 2022). To effectively use VRT in the agriculture industry, certain strategies may be helpful, particularly for land consolidation and bespoke hiring (Chaudhary et al., 2020). Along with the development of VRT, it is crucial to develop and support the demands and interests of small and marginal farmers (Chaudhary et al., 2020). Despite these challenges, VRT holds significant potential for improving efficiency and productivity in agriculture (Chaudhary et al., 2020; He 2022).

Conclusion

Given the high upfront costs and complex machinery involved, it's crucial to manage, calibrate, and maintain VRT effectively. Despite these challenges, the adoption of VRT can lead to a reduction in the total inputs required and boost crop yields by optimizing input utilization. This suggests that VRT holds substantial promise for future agricultural practices, especially if land consolidation becomes more common in developing nations like India. It is also vital to implement local interventions tailored to specific conditions and needs to promote the efficient and cost-effective use of VRT among farmers.

References

Abdel Rahman, M.A.E. "An overview of land degradation, desertification and sustainable land management using GIS and remote sensing applications." *Rendiconti Lincei. Scienze Fisiche e Naturali* 34, 3 (2023): 767-808.

Ahmed, Z., Gui, D., Qi, Z., Liu, Y., Liu, Y., Azmat, M. "Agricultural system modeling: current achievements, innovations, and future roadmap." *Arabian Journal of Geosciences* 15, 4 (2022): 363.

Bhattacharyya, R., Ghosh, B., Mishra, P., Mandal, B., Rao, C., Sarkar, D., et al. "Soil Degradation in India: Challenges and Potential Solutions." *Sustainability*, 7, 4 (2015): 3528-70.

Chaudhary, S., Negi, P.S., Singh, A., Prasad, R.K. "Variable rate application technology in India." 9, 9 (2020): 166-8.

Halawar, S.V. "The Trend Analysis of Major Food Grains in India." *International Journal of Current Microbiology and Applied Sciences* 8, 03 (2019): 353-60.

He, L. 2022. "Variable Rate Technologies for Precision Agriculture." *Encyclopedia of Smart Agriculture Technologies*. United States: Springer International Publishing. pp. 1-9.

Hossini, H., Karimi, H., Mustafa, Y.T., Al-Quraishi, A.M.F. 2022. "Role of Effective Factors on Soil Erosion and Land Degradation: A Review." *Environmental Degradation in Asia*. Germany: Springer International Publishing. pp. 221-35.

Jun, S., Park, J., Kim, J.Y. 2022. Digital Transformation Landscape in Asia and the Pacific: Aggravated Digital Divide and Widening Growth Gap. https://www.unescap.org/kp/2022/digital-transformation-landscape-asia-and-pacific-aggravated-digital-divide-and-widening [Last accessed February, 2024].

Juttu, R., Jogula, K., Priyadarshini, S., Bhoopal Reddy, S., Kumar Patra, P., Raju, B., et al. 2021. Strategies and Programs for Improved Nutrient Use Efficiency, Doubling Farmer's Income, and Sustainable Agriculture: Indian Context. *Technology in Agriculture*. IntechOpen. https://doi.org/10.5772/intechopen.98267 [Last accessed February, 2024].

Lampridi, M., Sørensen, C., Bochtis, D. "Agricultural Sustainability: A Review of Concepts and Methods." *Sustainability* 11, 18 (2019): 5120.

Liu, D.Y., Liu, Y.M., Zhang, W., Chen, X.P., Zou, C.Q. "Zinc Uptake, Translocation, and Remobilization in Winter Wheat as Affected by Soil Application of Zn Fertilizer." *Frontiers in Plant Science* 10 (2019).

MacPherson, J., Voglhuber-Slavinsky, A., Olbrisch, M., Schöbel, P., Dönitz, E., Mouratiadou, I., et al. "Future agricultural systems and the role of digitalization for achieving sustainability goals. A review." *Agronomy for Sustainable Development* 42, 4 (2022): 70.

Maja, M.M., Ayano, S.F. "The Impact of Population Growth on Natural Resources and Farmers' Capacity to Adapt to Climate Change in Low-Income Countries." *Earth Systems and Environment* 5, 2 (2021): 271-83.

Mistary, V.S. 2021. Agriculture inputs and their importance for productivity—an overview. 2021–2023. https://doi.org/10.36713/epra0813 [Last accessed February, 2024].

Mohapatra, T. 2022. "Precision Agriculture in India: A Perspective." *NAAS Newsletter* 22, 3 (2022): 1-4.

Padbhushan, R., Kumar, U., Sharma, S., Rana, D.S., Kumar, R., Kohli, A., et al. "Impact of Land-Use Changes on Soil Properties and Carbon Pools in India: A Meta-analysis." *Frontiers in Environmental Science* 9 (2022).

Rahate, O., Khandekar, Y. 2018. "Precision agriculture and its scenario in India." *Proceedings of the International Conference on Industrial Engineering and Operations Management* pp. 2637-40.

Randive, K., Raut, T., Jawadand, S. "An overview of the global fertilizer trends and India's position in 2020." *Mineral Economics* 34, 3 1(2021): 371-84.

Reddy, P.P. 2016. "Integrated Nutrient Management." *Sustainable Intensification of Crop Production.* Singapore: Springer. pp. 193-206.

Ritchie, H., Reay, D., Higgins, P. "Sustainable food security in India—Domestic production and macronutrient availability." *PLoS One* 13, 3 (2018): e0193766.

Rossetto, R., Ramos, N.P., de Matos Pires, R.C., Xavier, M.A., Cantarella, H., Guimarães de Andrade Landell, M. "Sustainability in Sugarcane Supply Chain in Brazil: Issues and Way Forward." *Sugar Tech* 24, 3 (2022): 941-66.

CEIC. 2021. Chemical Fertilizers: NPK Consumption: Per Hectare: Jammu and Kashmir: Nitrogen. https://www.ceicdata.com/en/india/chemical-fertilizers-nitrogen-phosphate-and-potash-npk-consumption-per-hectare-by-states/chemical-fertilizers-npk-consumption-per-hectare-jammu-and-kashmir-nitrogen.

Tayari, E., Jamshid, A.R., Goodarzi, H.R. "Role of GPS and GIS in precision agriculture." *Journal of Scientific Research and Development* 2, 3 (2015): 157-62.

Teague, R., Kreuter, U. "Managing Grazing to Restore Soil Health, Ecosystem Function, and Ecosystem Services." *Frontiers in Sustainable Food Systems* 4 (2020). https://doi.org/10.3389/fsufs.2020.534187.

Velten, S., Leventon, J., Jager, N., Newig, J. "What Is Sustainable Agriculture? A Systematic Review." *Sustainability* 7, 6 (2015): 7833-65.

Zamykal, D., Everingham, Y.L. 2009. "Sugarcane and Precision Agriculture: Quantifying Variability Is Only Half the Story—A Review." *Climate Change, Intercropping, Pest Control and Beneficial Microorganisms* Netherlands: Springer. pp. 189-218.

10

Soil Test Crop Response

Abstract

India's economy is heavily reliant on agriculture, with the majority of its population, around 70%, residing in rural areas. Fertilizers, despite being a costly input, play a crucial role in producing additional food grains to feed the growing population. The advent of high-yield crop varieties and the spread of multiple cropping systems under irrigation have led Indian farmers to recognize the significance of using balanced fertilizers based on soil test results, as the soil type greatly affects the fertilizer response. The use of balanced chemical fertilizers has become a necessity to meet the rising demand for food grains, pulses, oil crops, sugarcane, fruits, vegetables, fiber, forage, and grasses. However, it has been shown that improper fertilizer usage can degrade soil quality and impact nutrient use efficiency. As a result, managing soil fertility is critical for ensuring nutritional security and productivity, while also preserving soil health and sustainability. Fertilizers should be applied in the right quantity, from suitable sources, in the correct combination, at the right time, and using the right methods to maximize benefits, enhance nutrient use efficiency, and minimize nutrient loss. The recommended practices should be customized to accommodate soil variability.

Keywords: STCR, Soil test, Integrated Plant Nutrition System,Target yield

Introduction

In order to cater to the nutritional needs of a rapidly growing population, it is imperative to increase food grain production on the same land while maintaining the soil's productive capacity. This necessitates comprehensive research to establish a scientific foundation for enhancing and sustaining food production and soil productivity with minimal environmental impact (Rathore et al., 2010). Balanced nutrition extends beyond the application of nitrogen, phosphorous, and potassium in specific quantities through fertilizers. It entails ensuring that the soil contains the right proportions of available nutrients to meet the crop's requirements for optimal yield (Rathore et al., 2010). Achieving this balance requires replenishing the soil with the appropriate quantities of fertilizers or manures to replace the nutrients removed from the soil (Rathore et al., 2010). Soil tests provide valuable information about the available nutrient levels and any imbalances. Fertilizer recommendations aim to rectify these imbalances based on the crop's needs (Rathore et al., 2010). However, fertilizer recommendations based on a qualitative or semiquantitative approach do not yield the expected responses (Rathore et al., 2010). The All India Coordinated Research Project on Soil Test Crop Response (AICRP-STCR) has developed an inductive approach and refined fertilizer recommendations for different soil test values. Soil test response studies have used a targeted yield approach to establish a relationship between crop yield, soil test estimates, and fertilizer inputs. Research on soil biological parameters has shown that soil microbial biomass (MMB), dehydrogenase activity (DDE), and organic carbon (OC) were higher when nutrient application was based on STCR-based recommendations (Rathore et al., 2010). Farmers have seen significant agronomic and economic benefits from applying fertilizer nutrient doses in relation to soil tests. More recently, calibrations have been developed for an integrated supply of organic and fertilizers, taking into account nutrient contributions from organics, soil, and fertilizers (Rathore et al., 2010). Various centers have developed fertilizer prescription equations to simplify fertilizer usage based on soil test values. Applying fertilizers based on soil test results has helped achieve a higher response and benefit–cost ratio (BCR). It is evident that the STCR-based nutrient application approach has a clear advantage in increasing the nutrient response ratio compared to the generally recommended nutrient application dose. Demonstrations under the Tribal Sub Plan (TSP) in various states have shown the superiority of the STCR-based fertilizer recommendation for various crops compared to blanket recommendations and farmers' practices. The STCR approach is a leading scientific method for applying nutrients to crops based on soil test values (Reddy et al., 2022; Sachan et al., 2022). STCR equations are developed considering the contribution of nutrients from the soil, manures, and fertilizers (Reddy et al., 2022; Sachan et al., 2022). To achieve higher and sustainable yields while maintaining soil fertility and preventing environmental pollution, it is crucial to balance the nutrient requirements of crops with the available nutrient reserves in the soil. The STCR approach bases fertilizer recommendations on crop nutrient use efficiency and the nutrient contribution from the soil (Reddy et al., 2022; Sachan et al., 2022). This approach has been shown to be effective in various studies, demonstrating its potential for optimizing plant nutrient supply and yield (Reddy et al., 2022; Sachan et al., 2022).

Concept of STCR

The STCR approach aims to provide a basis for precise quantitative adjustments of fertilizer doses in response to varying soil test values under specific crop production levels, also known as target crop production levels. The STCR establishes a relationship between the soil test value and crop yield (Reddy et al., 2022; Sachan et al., 2022).These findings are then validated in subsequent field trials to support soil testing laboratories in their advisory role under specific soil, crop, and agroclimatic conditions. This approach has been shown to be effective in various studies, demonstrating its potential for optimizing plant nutrient supply and yield (Reddy et al., 2022; Sachan et al., 2022).

Objectives of STCR

- Establish a correlation between soil test values and crop response to fertilizer, in order to calibrate fertilizer recommendations based on soil test results
- Investigate the use of chemical fertilizers in combination with organic manures to enhance nutrient use efficiency
- Evaluate the applicability of various soil test methods under field conditions
- Develop a foundation for fertilizer recommendations for entire cropping sequences, based on initial soil test values
- Assess soil health under integrated nutrient management systems
- Formulate a soil test-based fertilizer prescription under an Integrated Plant Nutrition System (IPNS) for various crops and soils
- Determine the extent to which the use of organic manures in conjunction with chemical fertilizers can reduce crop fertilizer requirements
- Identify suitable soil test methods for field conditions
- Provide a scientific basis for improving and maintaining soil and crop productivity through a targeted yield approach of plant nutrient management, with minimal environmental degradation
- Establish the relationship between available soil test values in terms of nitrogen (N), phosphorus (P), and potassium (K), and the yield response to critical crops

- Formulate yield targeting equations for critical crops for fertilizer recommendations
- Determine the extent to which the conjunctive application of organic manure can reduce a crop's fertilizer needs

Scope of Soil Test Crop Response

The STCR approach has been used to develop recommendations for drip fertigation, which is a method of applying water and nutrients directly to the root zone of plants. This approach enhances nutrient use efficiency(Rangasamy et al., 2022). In the context of Vertisols in India, there has been an observed anomalous potassium response. This refers to unexpected crop responses to applied potassium in some Vertisols of India (Pandurang Gurav, 2018). Accounting for the contribution of nonexchangeable K, which is a form of potassium that is not readily available to plants, is crucial in these scenarios (Pandurang Gurav, 2018; Wanjari, 2014). The development of decision support systems (DSS) by integrating GPS/GIS (Global Positioning System/geographical information system)-based soil fertility maps with STCR prescription equations has been a significant advancement in precision agriculture (Devdas and Srivastava 2021; Mhaske et al., 2022). These systems use geospatial technologies to provide site-specific fertilizer recommendations, enhancing the efficiency of nutrient management (Devdas and Srivastava 2021; Mhaske et al., 2022). A universal extractant or method has been developed and calibrated for target yields of different crops (Hochmuth et al., 2014). This extractant is used in soil testing to determine the available nutrient content in the soil. The calibration process involves determining the crop nutrient requirement at different soil test values (Hochmuth et al., 2014). The STCR prescription equation has been used for the development of customized fertilizers (Ammal et al., 2020; Srijaya 2020). These fertilizers are tailored to the specific nutrient requirements of crops based on soil test values, enhancing nutrient use efficiency and crop productivity (Ammal et al., 2020; Srijaya 2020).

Approaches and Methods of STCR

The STCR experimentation primarily involves two approaches: the deductive approach and the inductive approach (Mukherjee and Tiwari 2020).The deductive approach operates on the premise that different levels of soil fertility are not expected to occur at a single location (Mukherjee and Tiwari 2020). Therefore, various sites are typically selected to represent different levels of soil fertility, and the inference is deduced and applied generally (Mukherjee and Tiwari 2020). On the other hand, the inductive approach to STCR field experimentation is based on the concept that variations in soil fertility are intentionally created at a single site within the same field (Mukherjee and Tiwari 2020). This approach minimizes the variability in soil type, management techniques used, and climatic conditions (Mukherjee and Tiwari 2020). STCR experiments are divided into two phases (Mukherjee and Tiwari 2020). The specifics of these phases would depend on the objectives of the study and the specific crop and soil conditions being investigated (Mukherjee and Tiwari 2020).

Development of Fertility Gradient

The first stage of STCR experimentation involves the selection of a field that is representative of the main soil type in the area (Mukherjee and Tiwari 2020). This field is divided into equal sections, each representing a different level of fertility. The first section does not receive any fertilizer, creating a baseline for comparison (Mukherjee and Tiwari 2020). In the subsequent sections, standard N, P, and K fertilizers are incrementally increased to create a gradient ranging from low to high fertility (Mukherjee and Tiwari 2020). Following the establishment of this fertility gradient, a short-duration exhaust crop is grown. During the growth of this crop, the applied nutrients undergo transformations in the soil due to interactions with plants and soil microbes (Mukherjee and Tiwari 2020). This process helps in understanding the nutrient dynamics in the soil and forms the basis for the next stages of the STCR experimentation (Mukherjee and Tiwari 2020).

Test Crop is Grown

The treatment combinations, which include N, P, and K, along with farmyard manure (FYM), are randomly distributed across each subplot, in addition to control groups (Hochmuth et al., 2014). A test crop, which requires soil test calibration, is then cultivated to maturity following standard agronomic procedures (Ahmadi et al., 2021; Beitlerová et al., 2021). Prior to the application of fertilizer, soil samples are collected from each subplot (Reijneveld et al., 2022). These samples are subsequently analyzed for available nutrients using established soil test methods (Hochmuth et al., 2014). Postharvest, the yield of grain and straw, as well as the total nutrient uptake, are determined (Ahmadi et al., 2021; Beitlerová et al., 2021). This process allows for a comprehensive understanding of the soil's nutrient-supplying capacity for the particular crop/plant during that growing season.

Development of Target Yield Equation

The concept of targeted yield in agriculture is rooted in Liebig's Law of Minimum (LBM), which posits that plant growth is limited by the nutrient in shortest supply in the soil, while all other nutrients are present in adequate amounts (Stine and Huybers 2017; Tang and Riley 2021). This law suggests that a specific quantity of soil nutrients is sufficient for a single yield of a particular percentage nutrient composition (Stine and Huybers 2017; Tang and Riley 2021). Among various fertilizer recommendation methods, the yield targeting method stands out. It not only indicates the soil test-based fertilizer dose but also the yield that the farmer can expect to achieve if proper agronomic techniques are employed to raise the crop (Tang and Riley, 2021). The "targeted yield" concept refers to a balance between the nutrients applied to the soil and the nutrients available in the soil (Tang and Riley 2021). This approach helps optimize nutrient use, improve crop yields, and promote sustainable farming practices (Tang and Riley 2021).

Soil Test-Based Approach of Fertilizer Management

GIS-based Fertility Maps

Fertility management using soil test-based methods is a proven strategy to enhance productivity in agricultural soils, which often exhibit high spatial variability due to the interplay of physical, chemical, and biological processes (Singh et al., 2021). However, the adoption of soil testing is limited in many developing countries due to challenges such as small-scale farming systems and lack of extensive soil testing infrastructure (Singh et al., 2021). GIS-based soil fertility mapping emerges as a promising solution (Biradar et al., 2020; Devdas and Srivastava 2021; Prabhavati et al., 2015). Utilizing GIS-based soil fertility maps as decision support tools for nutrient management not only promotes a rational approach over traditional farmer practices or blanket recommendations, but also eliminates the need for complex, individualized soil testing activities (Biradar et al., 2020; Devdas and Srivastava 2021; Prabhavati et al., 2015). Recent studies have shown promising results where GIS-based fertilizer maps have been used as nutrient decision support tools (Biradar et al., 2020; Devdas and Srivastava 2021; Prabhavati et al., 2015). These studies demonstrated that fertility maps, created using GIS and based on soil sampling at a 100 m^2 grid (one sample per hectare), helped estimate fertilizer requirements for farmers' fields, resulting in improved crop yield and profitability compared to soil test-based fertilizer recommendations for each field (Biradar et al., 2020; Devdas and Srivastava 2021; Prabhavati et al., 2015). The GIS approach, applied at a village scale at 76 locations within 543 landholdings, provided adequate fertilizer recommendation support for multiple field plots within the village to improve yield and profitability (Biradar et al., 2020; Devdas and Srivastava, 2021; Prabhavati et al., 2015). Several national agencies and Indian Agricultural Research Institute (ICAR) Institutes such as National Bureau of Soil Survey (NBSS) and Land Use Planning (LUP), Indian Institute of Science (IISS), Bhopal, and PDFSR have undertaken GIS-based fertility mapping initiatives at the state or regional scale. These initiatives are expected to facilitate the implementation of soil test-based fertilizer recommendations in larger domains without compromising on soil-testing services(Biradar et al., 2020; Devdas and Srivastava 2021; Prabhavati et al., 2015). In fact, the GIS-based approach is likely the only method that produces a visible impact of the soil test-based fertilizer recommendation within the existing soil-testing infrastructure (Biradar et al., 2020; Devdas and Srivastava 2021; Prabhavati et al., 2015).

Soil Test Crop Response Equations-based Fertilizer Recommendation

Soil testing is a crucial tool for determining the nutrient-sourcing capacity of soils, the crop's response to added nutrients, and the requirement of amendment (Liu et al., 2023; Montgomery and Biklé 2021; Tully and McAskill 2020). Soil test calibration, which establishes a relationship between the soil nutrient levels determined in the laboratory and the crop's reaction to fertilizer in the field, aids in balanced fertilization through the appropriate kind and rate of fertilizer (Liu et al., 2023; Reijneveld et al., 2022). LBM provides a theoretical basis and experimental evidence that N, P, and K work equally well (Stine and Huybers 2017; Tang and Riley 2021). This principle forms the foundation for the application of fertilizer for targeted yields, a concept first proposed by Truog in 1960 and later refined and expanded to other crops and soils by Ramamoorthy et al., (Bera et al., 2006; Liu et al., 2023).The yield targeting method of fertilizer recommendations is unique as it not only indicates the soil test-based fertilizer dose but also the level of yield a farmer can expect to achieve if proper agronomic techniques are employed in growing the crop (Tully and McAskill 2020). The essential basic data required for formulating fertilizer recommendations for the targeted yield concept include the nutrient requirement in kg/q of grain/other economic produce, the percent contribution from soil available nutrients, and the percent contribution from applied fertilizer nutrients (Tully and McAskill 2020). Fertilizer recommendations based on the yield targets concept have been tested in about 1800 frontline demonstrations conducted in farmers' fields in 15 states to validate the technology (Tully and McAskill 2020). Ready reckoners using the targeted yield equations have been developed for a wide range of crops on various soils (Tully and McAskill 2020).

Soil Test Crop Response—Understanding the Variability

Soil test crop response is viable technology to sustain higher crop productivity and soil quality for targeted yield indicates both soil test-based fertilizer dose and the level of yield that can be achieved with good agronomic practices. It works on objective to study the relationship between soil test values and response of yield and further derive yield targeting equations for adjusting fertilizer doses for different soil test values and response for targeted yield. Thus the concept of targeted yield concept comes under STCR. When the requirement of fertilizer for particular crop for its target yield is known than the requirement of fertilizer can be estimated by taking into account the efficiency of soil available nutrient pool and that of fertilizer requirement. In STCR approach, the target yield equation is obtained by:

$$F = T \times NR/CF - CS \times STV/CF - CO \times M/CF$$

Where,

F = Fertilizer dose of N, P_2O_5 or K_2O in kg ha

T = Yield target in q ha^{-1}

NR = Nutrient requirement of N, P_2O_5($P \times 2.29$) or K_2O ($K \times 1.21$) for 100 kg economic produce

CS = Contribution from soil nutrients in fraction

CF = Contribution from fertilizer nutrients in fraction

CO = Contribution from organic nutrients in fraction

STV = Soil available nutrients N, P_2O_5 ($P \times 2.29$) or K_2O ($K \times 1.21$) determined through soil analysis

M = Nutrient content in organic matter N, P_2O_5 ($P \times 2.29$) or K_2O ($K \times 1.21$) determined through organic matter analysis FYM.

Soil test crop response has developed algorithms of leaf color chart, SPAD, and fieldscout CM 1000 m values at three critical growth stages with yield in rice–wheat system; also developed fertilizer prescription equation for hitherto untouched secondary nutrient (sulfur) (Biswas et al., 2020).

Targeted Yield Approach

The targeted yield concept is a key component of the STCR approach (Haokip et al., 2021; Selvam et al., 2022). It refers to a specific quantity of nutrients that a crop must absorb to achieve a predetermined yield (Haokip et al., 2021; Selvam et al., 2022). This concept is crucial in striking a balance between fertilizing the crop and fertilizing the soil. Once the nutrient requirement for a given yield is known, the fertilizer requirement can be estimated. This estimation takes into account the efficiency of the available soil nutrient pool and the demand for fertilizer (Haokip et al., 2021; Selvam et al., 2022).

The STCR approach, which includes the targeted yield concept, has been popularized by the AICRP-STCR (Selvam et al., 2022). It involves a series of experiments that create a fertility gradient, grow an exhaust crop, grow a test crop with different treatments, and finally transform the data into a fertilizer prescription equation (Selvam et al., 2022). This equation uses three basic parameters: nutrient requirement by the crop, nutrient contribution from the soil, and nutrient contribution from fertilizer and organics (Selvam et al., 2022). The STCR approach provides a scientific basis for balanced fertilization and ensures a balance between applied nutrients and soil available nutrients. It has been shown to improve soil health and offer higher profit (Selvam et al., 2022). Fertilizer recommendations based on this concept are quantitative, precise, and meaningful as they involve both soil and plant analyses (Selvam et al., 2022).

Economic Analysis of Fertilizer Doses Associated with Different Yield Targets

A study evaluating the impact of NPK on crop yield and the BCR under NPK alone and Integrated Plant Nutrient System (IPNS) for 15 crops in agriculture and horticulture found that 66 crop × targets combinations resulted in a BCR between 1 and 2 in 35% of cases and a BCR between 2 and 3 in 62% of cases. In 3% of cases, the BCR was above 3 (Sun et al., 2022). Regardless of the crops, higher yields were observed with higher yield targets compared to lower targets, along with higher net return and BCR (Sun et al., 2022). As with yield, wherever three targets (low, mid, high) were tested, the difference in BCR between low and medium target levels was relatively higher than between medium and high target levels under NPK and IPNS (Sun et al., 2022).

Soil Test Crop Response-IPNS—3.5 t ha^{-1} has demonstrated its supremacy by registering higher grain yield of 3,455 kg ha^{-1} with a relatively higher response ratio of 12.57 kg kg^{-1} and BCR of 1.94 (Sun et al., 2022). STCR–IPNS—3.5 t ha^{-1} has recorded a yield increase of 43.1%, 21.2%, and 74.6% over blanket (100% RDF), blanket plus FYM 12.5 t ha^{-1}, and farmer's practice, respectively and the increase in response ratio was 4.28, 1.18, and 5.27 kg kg^{-1}, respectively (Sun et al., 2022). Therefore, for achieving higher grain yield, nutrient uptake, response ratio, and BCR in pearl millet (var. CO 10) on a mixed black calcareous soil (Periyanaickenpalayam soil series—Vertic Ustropept), STCR-IPNS-based fertilizer prescriptions can be followed (Sun et al., 2022).

Farmers can choose their desired yield targets based on their investment capacity and availability of (organic) manures, and they would generally benefit from adopting a suitable IPNS package (Sun et al., 2022). Organic manures not only contribute nutrients to the soil but also improve soil physical conditions (Sun et al., 2022). Soil test-based recommendations currently have a better chance of being valid if they only include fertilizers as opposed to IPNS because there are several nutrient issues that need to be addressed, as shown using STCR data from Andhra Pradesh (Sun et al., 2022).

In an effort to generate spatial fertilizer recommendation maps, a joint venture of the IISS, and the NBSS and LUP, Nagpur, utilized validated fertilizer adjustment equations and GIS to generate a district-level soil fertility index for 10 Indian districts (Sun et al., 2022). This index can be used to create balanced fertilizer recommendations for specified crops for the entire district, based on the district's average soil fertility status (Sun et al., 2022). Based on the index values for nitrogen, phosphorus and potassium (N, P, and K) for ten states, geo-referenced district-wise soil fertility maps were prepared (Sun et al., 2022). Adequate limits for targeted yields have been defined, and the recommendations in form equations for target yields have been linked to the fertility maps (Sun et al., 2022). Levels may vary depending on the availability of information. This recommendation system can be used for a variety of targeted yield applications in various districts of states. It can also be applied up to field level if the farmer is aware of their fertility status and yield target (Sun et al., 2022). These maps can be updated periodically based on their soil test results database. It can be narrowed down to block/village level depending on the availability of information (Sun et al., 2022). A case study was conducted at Honnavalli microwatershed in Hassan District, Karnataka, to prepare digital maps for site-specific fertilizer recommendations for the two major crops (Paddy and Maize) using the STCR approach (Sun et al., 2022). The results of the study indicated that the recommended dose of N fertilizer for paddy is 148, 111, 82, 54, 26, −8, −67 and −108 kg ha^{-1} for the soil available N status <200, 200–250, 250–280, 280–325, 325–375, 375–425, 425–500, and >500 kg, respectively. Similarly, for maize it is 290, 286, 283, 281, 279, 276, 268 and 259 kg ha^{-1} for <250, 250–275, 275–300, 300–325,

325–350, 350–400, 400–500 and >500 kg ha^{-1}, respectively (Sun et al., 2022). Therefore, the study has implications on reducing consumption of fertilizers (24–45% for N; 12–15% for P and 8–32% for K), thereby reducing the cost of cultivation besides achieving higher nutrient use efficiency (Sun et al., 2022).

Conclusion

The STCR approach emerges as a crucial and scientifically sound method for optimizing fertilization strategies in the context of India's agriculture, where most of the population relies on rural areas and the economy heavily depends on agriculture. The imperative to increase food grain production on the same land while preserving soil productivity necessitates a comprehensive and targeted approach to nutrient management. The STCR approach, rooted in establishing a quantitative relationship between soil test values and crop response to fertilizers, has proven to be effective in various studies. By calibrating fertilizer recommendations based on soil test results, the approach ensures a balance between applied nutrients and soil-available nutrients, promoting sustainable farming practices. The targeted yield concept, a key component of the STCR approach, provides a nuanced understanding of the specific nutrient quantities required for predetermined crop yields. This allows for precise and meaningful fertilizer recommendations, optimizing nutrient use efficiency and crop productivity.

References

Ahmadi, A., Emami, M., Daccache, A., He, L. (2021). "Soil Properties Prediction for Precision Agriculture Using Visible and Near-Infrared Spectroscopy: A Systematic Review and Meta-Analysis." *Agronomy* 11, 3 (2021): 433.

Ammal, U.B., Sankar, R., Coumaravel, K., Dey, P. "Fertilizer Prescriptions for Rice Based on STCR–IPNS." *International Journal of Current Microbiology and Applied Sciences* 9, 10 (2020): 2517-22.

Beitlerová, H., Lenz, J., Devátý, J., Mistr, M., Kapička, J., Buchholz, A., et al. "Improved calibration of the Green–Ampt infiltration module in the EROSION-2D/3D model using a rainfall-runoff experiment database." *Soil* 7, 1(2021): 241-53.

Bera, R., Seal, A., Bhattacharyya, P., Das, T.H., Sarkar, D., Kangjoo, K. "Targeted yield concept and a framework of fertilizer recommendation in irrigated rice domains of subtropical India." *Journal of Zhejiang University Science B* 7, 12 (2006): 963-8.

Biradar, B., HM, J., S, C., KN, G., Sannagoudar, M.S., AS, P., KN, P. "Assessment of soil fertility through GIS techniques and thematic mapping in micro-watershed of Hassan, Karnataka." *Journal of Pharmacognosy and Phytochemistry* 9, 4 (2020): 3218-28.

Devdas, D., Srivastava, L.K. "Soil fertility mapping by geographical information system (GIS) in different blocks of Gariaband district, Chhattisgarh." *Journal of Pharmacognosy and Phytochemistry* 10, 2 (2021): 328-38.

Haokip, I.C., Das, H., Dey, P., Kumar, D. "Targeted yield approach based fertilizer recommendations." *Harit Dhara* 4, 1 (2021): 1-3.

Hochmuth, G., Mylavarapu, R., Hanlon, E. "Developing a Soil Test Extractant: The Correlation and Calibration Processes." *EDIS* 8 (2014): 1-4.

Liu, J., Cai, H., Chen, S., Pi, J., Zhao, L. "A Review on Soil Nitrogen Sensing Technologies: Challenges, Progress and Perspectives." *Agriculture* 13, 4 (2023): 743.

Mhaske, G.D., Wadekar, A.V, Kajale, R.S., Raundal, A.V. "GPS-GIS based Soil Fertility Status of Jam River Basin , Nashik and Ahmednagar District (M.S.), India." *International Journal of Advance and Applied Research* 9, 6 (2022): 695-703.

Montgomery, D.R., Biklé, A. "Soil Health and Nutrient Density: Beyond Organic vs. Conventional Farming." *Frontiers in Sustainable Food Systems* 5 (2021).

Mukherjee, A., Tiwari, M. "Utilisation of Quantum Dots in Food Industry." *Agri Mirror: Future India* 1, 5 (2020): 4-7. Available from: https://aiasa.org.in/wp-content/uploads/2020/09/2.pdf.

Pandurang Gurav, P. "A review on soil potassium scenario in vertisols of India." *Open Access Journal of Science*, 2, 1 (2018).

Prabhavati, K., Dasog, G.S., Patil, P.L., Sahrawat, K.L., Wani, S.P. "Soil Fertility Mapping using GIS in Three Agro-climatic Zones of Belgaum District, Karnataka." *Journal of the Indian Society of Soil Science* 63, 2 (2015): 173.

Rangasamy, S., Subramaniyam, M., Stephen, P.K., Dey, P. 2022. "Drip Fertigation with Fertilizer Prescription Through STCR—IPNS—A Way Forward Towards Climate Change Mitigation." *Advanced Modelling and Innovations in Water Resources Engineering. Select Proceedings of AMIWRE 2021*. Edited by Rao, C.M., Patra, K.C., Jhajharia, D., Kumari S. Singapore: Springer Nature Singapore; pp. 749-57). https://doi.org/10.1007/978-981-16-4629-4_52

Rathore, R., Rathore, S., Sharma, N., Bidwai, A. *Pharmacology Online* 3 (2010): 291-7.

Reddy, K.J., Chandrashekara, C.P., Biradar, D.P., Rajakumar, G.R., Kanannavar, P.S. "Effect of Soil Test Crop Response (STCR) and STCR –Integrated Plant Nutrition System (IPNS) Based Fertilizer Prescription Equations on Maize Productivity, Economics and Soil Fertility Status in Vertisols of Northern Karnataka." *International Journal of Plant & Soil Science* (2022): 1-8.

Reijneveld, J.A., van Oostrum, M.J., Brolsma, K.M., Fletcher, D., Oenema, O. "Empower Innovations in Routine Soil Testing." *Agronomy* 12, 1 (2022): 191.

Sachan, R., Thomas, T., Pandey, H.P., Tiwari, A., Kumar, A., Yadav, P., et al. (2022). "Effect of STCR-IPNS based nutrient application on soil health, yield, nutrient content and uptake of mustard (Brassica Juncea l.) In Eastern Plain Zone of Uttar Pradesh, India." *International Journal of Plant & Soil Science* 2022: 248-58.

Selvam, R., Santhi, R., Maragatham, S., Gopalakrishnan, M., Rejeswari, R. "Targeted yield model-based balanced nutrient recommendation for barnyard millet on inceptisol of Tamil Nadu." *Indian Journal of Agricultural Research* 56, 6 (2022): 638-45.

Singh, V.K., Gautam, P., Nanda, G., Dhaliwal, S.S., Pramanick, B., Meena, S.S., et al. "Soil Test Based Fertilizer Application Improves Productivity, Profitability and Nutrient Use Efficiency of Rice (oryza sativa l.) Under Direct Seeded Condition." *Agronomy* 11, 9 (2021): 1756.

Srijaya, T. "Fertilizer Prescription Equation Demonstration for Targeted Yield of Rice in Farmer Fields of Telangana State." *Chemical Science Review and Letters* 9, 36 (2020): 916-21.

Stine, A.R., Huybers, P. "Implications of Liebig's Law of the Minimum for Tree-Ring Reconstructions of Climate." *Environmental Research Letters* 12, 11 (2017): 114018.

Sun, J., Luo, H., Jiang, Y., Wang, L., Xiao, C., Weng, L. "Influence of Nutrient (NPK) Factors on Growth, and Pharmacodynamic Component Biosynthesis of Atractylodes chinensis: An Insight on Acetyl-CoA Carboxylase (ACC), 3-Hydroxy-3-Methylglutaryl-CoA Reductase (HMGR), and Farnesyl Pyrophosphate Synthase (FPPS) Si." *Frontiers in Plant Science* 13 (2022).

Tang, J., Riley, W.J. "Finding Liebig's Law of the Minimum." *Ecological Applications* 31, 8 (2021).

Tully, K.L., McAskill, C. "Promoting Soil Health in Organically Managed Systems: A Review." *Organic Agriculture* 10, 3 (2020): 339-58.

Wanjari R.H., M.V. Singh, P.K. Ghosh. Sustainable Yield Index: An Approach to Evaluate the Sustainability of Long-Term Intensive Cropping Systems in India. *Journal of Sustainable Agriculture* 24, 4 (2004): 39-56.

11

Project Formulation Related to Precision Farming

Abstract

Precision farming is based on the idea of site-specific management, which aims to adjust field management according to specific soil conditions to optimize production and minimize environmental damage. However, site-specific management is only relevant if agronomists can provide accurate site-specific advice. This requires detailed information about specific soil conditions, particularly soil characteristics and soil behavior. While remote-sensing (RS) data is one of the main sources for the required information, it has been established that essential data for precision farming can hardly be obtained from low or medium spatial resolution satellite data. At the same time, high- and very high-resolution data are far from cost-effective. In Mediterranean agronomical conditions, which contain significant heterogeneity, small particle size, and low per capita income, the problem is magnified. To overcome these difficulties, fast and inexpensive RS data must be available to farmers and agronomists at any time during the season. In this study, a new technique was developed to respond to this need, using a ground RS tool [a red-NIR (near-infrared) camera], and a potential application of this technique is examined. The technique involves the retrieval of photographed plots taken from the field under study; the camera's accompanying software can calculate vegetation indices. The combination of the plots index values with the plot's spatial data (taken from a grid on the ground) results in an attribute table, which is then processed by geographic information system (GIS) (geoinformatics tools) (interpolation techniques) to produce thematic maps. These maps can then be used to track spatial or temporal variability of the wheat crop field. The usefulness of this method lies in the ability of farmers to monitor the state of their fields at any time during the cultivation season, identify problematic areas, and apply inputs according to the needs of the field area and time specifically.

Keywords: Precision farming, red-NIR camera, GIS, spatial pattern

Introduction

Precision farming, also known as precision agriculture, prescription farming, and site-specific farming, is grounded on the principle of managing fields based on their heterogeneity (Nowak 2021). In the past, crop management decisions were made for the entire field, treating it as a homogenous unit, despite significant variations in soil characteristics, weed infestations, and plant growth within a field (Nowak 2021). Precision agriculture integrates modern technology with past experience to manage farming on a grid-by-grid basis (Nowak 2021). Instead of managing an entire field based on an average hypothetical condition, precision farming acknowledges site-specific variations within each field and adjusts management decisions accordingly (Nowak 2021). This has been made possible due to the emergence of technology that allows soil management to be altered automatically as machinery traverses the field (Nowak 2021). Precision farming aims at maximizing production and minimizing environmental damage (Nowak 2021). Fields are the management units a farmer has to deal with when managing his land. Management decisions are based on either the measured or perceived conditions of the field for soils and crops at any given time, both now and in the near future (Nowak 2021). Management decisions to be made by the farmer have different dimensions (Nowak 2021). First, the strategic one: he has to make major decisions about the future direction of his farm; for example, which crop rotation to follow, how large his herd of cows will be, and what type of farm management he will pursue (Nowak 2021). Will it follow the principles of organic, biological, ecological, or conventional farming? Will it be labor- and capital-intensive or based on part-time labor elsewhere, etc. (Nowak 2021). Strategic decisions reflect his vision regarding the sustainability of his farming system in terms of yield, profitability, risk assessment, environmental quality, and social acceptability (Nowak 2021). Then, decisions have a tactical dimension; for example, the choice of crop varieties or the amount of chemical fertilizers to be used. Finally, there are operational decisions: when and how much to fertilize, when to apply biocides, when and how to till, etc. Precision farming deals with all types of decisions; however, its main emphasis is on operational decisions: timing and type of tillage, seeding and planting, fertilization, pest management, and timing of harvest (Nowak 2021). However, site-specific management is only relevant if agronomists can provide accurate site-specific advice (Nowak 2021). This requires specific information on specific soil conditions, particularly soil characteristics and soil behavior (Nowak 2021). One of the major sources for the acquisition of this information is remote-sensing (RS) data. RS is the science of obtaining and analyzing data about the environment by sensors without their being in physical contact with those environments. For project formulation, we should have theory knowledge of precision farming its rationale, its methodology, merits and demerits so that we are able to frame the project according to format.

Precision farming starts at the crop growth cycle, that consists of five basic steps:

1. *Precision soil preparation*: Precision soil preparation is a crucial step in enhancing the efficiency of limited natural resources in a sustainable manner (Rayhan Shaheb et al., 2022). The integration of information technology and agricultural science, known as precision agriculture, utilizes modern tools to improve economically and environmentally sustainable crop production (Nowak 2021; Rayhan Shaheb et al., 2022). An integrated crop management system aims to match the type and amount of inputs with the actual crop needs for small areas within a farm field (Rayhan Shaheb et al., 2022)
2. *Precision seeding*: Precision seeding involves placing the desired number of seeds at a specific depth and spacing (Ji et al., 2021). This method is particularly beneficial for vegetable growers as it offers more advantages over conventional dribble or multi-seed drop-plate seeding systems (Ji et al., 2021)
3. *Precision crop management*: Also known as precision crop load management (PCLM), precision crop management is a scientific approach used to calculate the optimal crop loads of fruit crops to maximize year-over-year profits (Nowak 2021; Rayhan Shaheb et al., 2022)
4. *Precision harvesting*: Precision harvesting is a modern agricultural technique that employs various cameras and sensors to determine the optimal time for cultivation and to maximize yield (Nowak 2021). These sensors enhance crop efficiency and impact crop returns by monitoring humidity, temperature, crops, and soil (Nowak 2021)
5. *Knowledge analysis and evaluation*: The process of knowledge analysis and evaluation involves determining and defining the methods, roles, and responsibilities for collecting and analyzing data using appropriate statistical and nonstatistical techniques (Weißhuhn et al., 2018). This procedure is crucial in the field of agriculture for making informed decisions and improving farming practices

Rationale

Precision farming commonly utilizes data sources such as yield maps produced by yield monitors installed on harvest machines, real-time soil analysis through sensors installed on the tractor, and more traditional soil and plant tissue analyses conducted in the laboratory (Misara et al., 2022; Nowak 2021). All the collected data are processed through geographical information systems (GIS), which represent a dynamic set of tools for the collection, storage, retrieval, transformation, and output of spatial data on the environment to meet certain specialized demands (Onyango et al., 2021). For the application of inputs according to the recorded variability in the parcel, variable-rate technology (VRT) is used (He 2022). Mediterranean agronomic conditions are characterized by significant heterogeneity, small parcel size, and low per capita income, while the wheat crop is typically low labor- and capital-intensive (Bill et al., 2022). This means that Mediterranean farmers can only afford small investments when tracing variability on a wheat parcel (Bill et al., 2022). In terms of RS, the current service provided by low-cost commercial satellites is inadequate (Mathenge et al., 2022). A past survey showed that <25% of the essential parameters of the wheat crop can be obtained using existing satellite data (Bill et al., 2022). This can be attributed to the low spatial and temporal resolution of the inexpensive satellite data, while very high-resolution data proved to be far from cost-effective (Bill et al., 2022; Ghosh and Kumpatla 2022). To overcome this problem, fast and cheap data should be made available to farmers during the cultivation season, especially in the early growth stages of the crop (Ghosh and Kumpatla 2022). Recently, in response to this need, an integrated GIS-GPS (Global Positioning System) system was developed for locating point, line, or polygon in the field, in real time (He 2022). This system employs RS devices on the ground (R-NIR camera and hyperspectral radiometer) to record variability in the field (He 2022). Also, spectral models were developed for use in biomass and canopy cover percent estimation of the wheat crop in the tillering stage. Only photographed plots taken with a specific red-NIR (near-infrared) camera or sampling biomass measurements of the cultivated area are needed to use these models (Thompson et al., 2019).

Methodology

The process of growing crops involves the use of RS and Geographic Information Systems (GIS), particularly when integrating ground Normalized Difference Vegetation Index (NDVI) measurements with interpolation methods (Amini et al., 2019; Mathenge et al., 2022). This technique employs a special red-NIR camera capable of calculating vegetation indices and a GIS software package, and is carried out in three stages:

- *Data collection*: The first stage begins in the laboratory and continues in the field. A sampling grid on the wheat parcel is established by measuring the desired distances from the edges. A number of plots are then photographed throughout the grid using a systematic sampling method with a special camera, the ADC-R-NIR camera (Mathenge et al., 2022). This camera has a spectral range of 600–1,000 nm, an image resolution of 496 × 365 pixels, and a color resolution of 8 bits/bands. It supports various vegetation indices including NDVI, IPVI, and SAVI (Mathenge et al., 2022)
- *Calibration and NDVI calculation*: After the sampling is completed, each image undergoes a calibration procedure. In each calibrated image, the NDVI is automatically calculated (Mathenge et al., 2022). If, for example, 18 plots have been collected, 18 NDVI values must be calculated (one per plot) (Mathenge et al., 2022)
- *Creation of thematic maps*: The creation of the thematic maps using interpolation techniques is carried out automatically by GIS software packages (Mathenge et al., 2022). In this study, ESRI ArcGIS was used. Interpolation is the process whereby a value is estimated at a location based on available samples (Amini et al., 2019). There are two groups of interpolation techniques: deterministic and geostatistical (Amini et al., 2019). Deterministic interpolation techniques create surfaces from measured points, based on either the extent of similarity [e.g., inverse distance weighted (IDW)] or the degree of smoothing (e.g., radial basis functions). Geostatistical interpolation techniques (e.g., Kriging) utilize the statistical properties of the measured points (Amini et al., 2019)

The necessary equipment for this process includes an red-NIR camera equipped with software capable of calculating vegetation indices, a personal computer, and a GIS software package for geographic analysis (Tsuchikawa et al., 2022). Additional materials required for the practical work include guides for the field grid, such as flags, and a printer for the printouts of the thematic maps or statistics (Tsuchikawa et al., 2022).The obtained NDVI thematic charts can be utilized to track variability in crop growth and subsequently in the operation of appropriate inputs (e.g., nitrogen fertilization, irrigation), according to the site-specific requirements (Tsuchikawa et al., 2022). The charts are produced with predetermined detail (classes or situations of crop growth), which depends on the sensitivity level at which the farmer intends to manage his crop (Tsuchikawa et al., 2022). Two strategies can be adopted by the farmer or the agriculturist:

1. The first strategy assumes that the degree of labor and capital intensity is not predefined by the farmer and that he intends to follow what the soil or plant analysis dictates (Tsuchikawa et al., 2022). Soil analysis can provide results regarding the state of the nutrients in the soil (e.g., nitrogen, phosphorus, potassium), whereas plant tissue analysis can directly indicate the state of nutrients in the tissues (e.g., nitrogen) (Tsuchikawa et al., 2022)

2. The second strategy assumes that the degree of labor and capital intensity is given; in this case, soil analysis is not performed and the thematic chart can be used for the appropriate distribution of the given treatment across the parcel, according to the relative site-specific requirements (Tsuchikawa et al., 2022)

- After the conditions of the variable parameters, such as nitrogen fertilization, have been defined by the soil or plant analysis or the calculation distribution procedure, the operation stage follows (Tsuchikawa et al., 2022). The thematic charts should typically be used by special VRT equipment at this stage (He 2022). However, this type of equipment is relatively expensive in terms of investing in a wheat crop (Tsuchikawa et al., 2022). Therefore, this study proposes a method to overcome the need for the aforementioned machines (Tsuchikawa et al., 2022). A prerequisite for this alternative method to precisely apply the inputs is the scaling up of the thematic charts, which results in new thematic charts in a coarser resolution (Tsuchikawa et al., 2022). For example, if the swath of the operation is 20 meters, scaling up to a resolution of a 20 m pixel can be chosen (Tsuchikawa et al., 2022)

Consequently, the tractor with the mounted equipment (e.g., a conventional pesticide distributor), instead of using a VRT system, could select its speed among a number of predefined rates according to the state of the specific site's requirements. In this way, the variable speed rates of the tractor compensate for the nonvariable (Tsuchikawa et al., 2022).

Merits

- *GPS in agriculture*: GPS technology has revolutionized modern farming and has become an essential tool in precision agriculture. It allows for accurate mapping and surveying of fields, which can help in planning and optimizing agricultural practices (Mathenge et al., 2022; Tayari et al., 2015)

- *Mapping yield and soil traits*: Advanced technologies are being used to map yield and soil traits in agriculture. This includes the use of GIS and RS applications. These technologies can help in crop yield estimation, soil fertility assessment, and precision agriculture (Mathenge et al., 2022; Suruliandi et al., 2021)
- *Subdividing nonuniform fields*: While there's limited research specifically on subdividing nonuniform fields in agriculture, the concept of precision agriculture often involves managing smaller, variable-rate zones within a field. This approach can lead to more efficient use of resources and improved crop performance (Tayari et al., 2015)
- *Resource management and reducing wastage*: Sustainable agricultural practices aim to improve resource management and reduce wastage. This includes optimizing the use of agrochemicals, minimizing food loss and waste, and implementing practices that enhance soil health and biodiversity (Agarwal et al., 2021; LaCanne and Lundgren 2018; Mitra et al., 2021)
- *Minimizing environmental risks*: One of the major environmental risks in agriculture is nitrate leaching and groundwater contamination. Optimizing the use of agrochemicals, particularly fertilizers, can help minimize these risks. Sustainable agricultural practices, including the use of organic fertilizers and precision agriculture techniques, can play a crucial role in reducing nitrate leaching and protecting groundwater resources (Gupta et al., 2020; Pérez-Ruiz et al., 2011)

Demerits

- *Development of techniques*: Precision agriculture and other advanced farming techniques are still under development (Carlisle et al., 2019; Khanal and Omobitan 2020). Therefore, it is crucial to seek expert advice before making significant decisions. This ensures that farmers and agricultural professionals make informed choices that align with the latest research and technology (Carlisle et al., 2019; Khanal and Omobitan 2020).
- *Initial capital expenses*: The initial capital expenses for implementing advanced farming techniques can be high (Carlisle et al., 2019; Khanal and Omobitan 2020). This is particularly true for precision agriculture, which requires investment in sophisticated equipment and technology (Carlisle et al., 2019; Khanal and Omobitan 2020). Therefore, it should be viewed as a long-term investment.

- *Data collection and analysis*: Implementing advanced farming techniques often requires several years of data collection to fully implement the system (Carlisle et al., 2019; Khanal and Omobitan 2020). This data, which can include information on soil characteristics, crop yields, and weather patterns, is crucial for optimizing farming practices
- *Demanding work*: Collecting and analyzing data for precision agriculture or other advanced farming techniques can be extremely demanding (Carlisle et al., 2019; Khanal and Omobitan 2020). It requires a significant amount of time and effort as well as expertise in data analysis and interpretation. Despite these challenges, the potential benefits of these techniques, such as increased crop yields and improved resource efficiency, make them a worthwhile investment (Carlisle et al., 2019; Khanal and Omobitan 2020)

Different Agencies for Submission of Project

Ministry of Environment and Forests, Forests Research Institute , Department of Science and Technology (DST), Department of Atomic Energy (DAE), Department of Biotechnology (DBT), Department of Education (DOEd), Department of Food Processing Industries Government of India, Department of Nonconventional Energy Sources (DNES), National Information System for Science and Technology (NISSAT), Technology Absorption and Adaptation Scheme, Department of Space (DOS), Science and Engineering Research Council (SERC), intensification of research in high priority areas, science and society-related programs, Science and Technology Application for Rural Development (STARD), Science and Technology for Weaker Sections (STAWS), Women Component Plan (WCP) science and society-related programs, utilization of scientific expertise of retired scientists science and society related programs, special component plan, science and technology communication and popularization program, Natural Resources Data Management System (NRDMS), Instrument Development Programme (IDP), Opportunities for Young Scientists, Science and Technology indicator and Manpower Studies, Consumer Protection through Science and Technology, etc.

Project Formulation Format

It can vary for different agencies one example is given as follows (Table 1):

Table 1: Summary of the project.

1.	Title of proposal	
2.	Name of the principal investigator (PI)	
3.	Name of the Co-PI	
4.	Location	
5.	Mailing address	
6.	Tel. No. and Fax No.	
7.	E-mail	
8.	Duration	
9.	Budget	

Title of the project: Precision agriculture for higher productivity, profitability and nutritional security under subtropical conditions of Jammu.

Project summary: Precision farming coming under two-dimensional farming is confronting with major problems of rapid urbanization, climate change, water scarcity, and decreasing land availability/capita. Hence, it is unable to meet the nutritional food security of exponentially increasing population. Thus there is a crying need to switch to an alternative method of precision farming that aims to study the feasibility of suitable precision practices for higher productivity, profitability, and nutritional security under subtropical zones.

Introduction and rationale: The present rates of crop growth yield and production under the present scenario of climate change and fragmented landholding of 1.15 ha. Thus to meet the food and nutritional security under shrinking resources the researchers, planners have to designed the precision farms. Thus research initiatives should be taken in precision farming for nutritional security under shrinking land resources.

Objectives:

- To evaluate the productivity and profitability of under precision farming
- To evaluate the relative economics of different treatments
- To impart trainings cum demonstration to the urban dwellers/farmers/ unemployed youth with respect to precision farming under subtropical conditions

Review of Research Conducted/Being Conducted on the Subject

International Work (250 words)

The major constraint of precision farming is its start-up cost due to adoption of conventional commercial glasshouse agriculture, but it provided greater crop yield per unit of land (Beacham 2019). Precision is more suitable than traditional farming (Sills 2018). Corvalan et al. (2005) compared the precision farming system (VFS) with conventional horizontal system (HHS) and found VFS produced more crops per unit of growing floor area when compared with the HHS, but light distribution and shoot fresh weight decreased significantly in the VFS from top to base.

National (250 Words)

The integration of precision farming with hydroponics mitigates the problem of food scarcity for the future but required more dedication of time and workforce over traditional farming. (Kumar 2019). According to ICAR (2016) report vertical aeroponic farm of 400 sqft area has the distinct advantage of producing the same quantum of better-quality produce than a 1,600 sqft soil-based farm.

Technical Program

The treatment details are given in Table 2.

Table 2: Treatment details.

Treatments	
Precision farming treatments:	
S_1	
S_2	
S_3	
S_4	
Cropping sequences:	
C1	
C2	
C3	

Activity milestones in first year:

- Selection of suitable site
- Construction of precision structures as per technical program
- Procurement of various inputs as per technical program

Second year:

- Sowing/transplanting of different types of crops as per technical program
- Application of soluble fertilizers and irrigation as per the crop requirement
- Compilation of results

Third year:

- Sowing/transplanting of different types of crops as per technical program
- Compilation of results and submission of report

Duration: 3 years

Approaches for sustainability:

- Achieving nutritional security by providing quality food round the year
- Saving of resources, viz., land, human, water, fossil fuels, etc.
- Earning profits through diversified and sustainable round the year quality crop enterprises

Institutional competence in the area: Our institute for improving nutritional and livelihood security.

Likely beneficiaries: Farmers with low landholdings, will provide quality nutritional security/capita

Budget summary: It is given in Table 3.

Table 3: Budget summary.

S. No.	Description of items	Years				Total cost (lakh)	Justification
A	Nonrecurring	Quantity (no)	First year	Second year	Third year		
	Subtotal of A	–	–	–	–		
B.	Recurring cost						
	Subtotal of B						
C.	Grand total (A + B)						
	Institutional charges @10% of nonrecurring cost						
D.	Total budget required						

Certificate

- The scales of pay, allowances, etc., are those admissible to persons of corresponding status employed in the university
- The university will provide the basic facility for undertaking the proposed project to the investigator(s)

Signature of P.I.

Expected outcome:

- The study shall provide best possible solution to food security by providing safe food and nutritional security.
- The study shall ensure the optimum utilization of farm inputs
- It shall be helpful for developing entrepreneurship among the unemployed youth for precision farming

Conclusion

For project formulations current problems and futuristic scenarios should be taken in to consideration along with capacity building in farmers. Right agency or organization should be approached on the basis of theme suitable to them. Innovative ideas are always encouraged so framed projects should be innovative.

References

Agarwal, M., Agarwal, S., Ahmad, S., Singh, R., Jayahari, K.M. 2021. Food Loss and Waste in India: The Knowns and the Unknowns. World Resources Institute. https://doi.org/10.46830/wriwp.20.00106 [Last accessed February, 2024].

Amini, M.A., Torkan, G., Eslamian, S., Zareian, M.J., Adamowski, J.F. "Analysis of deterministic and geostatistical interpolation techniques for mapping meteorological variables at large watershed scales." *Acta Geophysica* 67, 1 (2019): 191-203.

Bill, R., Nash, E., Grenzdörffer, G., Wiebensohn, J. "Geographic Information Systems in Agriculture." *Springer Handbook of Geographic Information*. Switzerland: Springer International Publishing. pp. 659-84.

Carlisle, L., de Wit, M.M., DeLonge, M.S., Calo, A., Getz, C., Ory, J., et al. "Securing the future of US agriculture: The case for investing in new entry sustainable farmers." *Elementa: Science of the Anthropocene* 7, 17 (2019).

Ghosh P., Kumpatla, S.P. 2022. "GIS Applications in Agriculture." *Geographic Information Systems and Applications in Coastal Studies*. IntechOpen. https://doi.org/10.5772/intechopen.104786 [Last accessed February, 2024].

Gupta, P.K., Kumari, B., Gupta, S.K., Kumar, D. "Nitrate-leaching and groundwater vulnerability mapping in North Bihar, India." *Sustainable Water Resources Management* 6, 3 (2020): 48.

He, L. 2022. "Variable Rate Technologies for Precision Agriculture." *Encyclopedia of Smart Agriculture Technologies*. Edited by Zhang, Q. United States: Springer International Publishing. pp. 1-9.

Ji, J., Sang, Y., He, Z., Jin, X., Wang, S. "Designing an intelligent monitoring system for corn seeding by machine vision and Genetic Algorithm-optimized Back Propagation algorithm under precision positioning." *PLOS One 16*, 7 (2021): e0254544.

Khanal, A.R., Omobitan, O. "Rural Finance, Capital Constrained Small Farms, and Financial Performance: Findings from a Primary Survey." *Journal of Agricultural and Applied Economics* 52, 2 (2020): 288-307.

LaCanne, C.E., Lundgren, J.G. "Regenerative agriculture: merging farming and natural resource conservation profitably." *PeerJ* 6 (2018): e4428.

Mathenge, M., Sonneveld, B.G.J.S., Broerse, J.E.W. "Application of GIS in Agriculture in Promoting Evidence-Informed Decision Making for Improving Agriculture Sustainability: A Systematic Review." *Sustainability* 14, 16 (2022): 9974.

Misara, R., Verma, D., Mishra, N., Rai, S.K., Mishra, S. "Twenty-two years of precision agriculture: a bibliometric review." *Precision Agriculture* 23, 6 (2022): 2135-58.

Mitra, B., Chowdhury, A.R., Dey, P., Hazra, K.K., Sinha, A.K., Hossain, A., et al. 2021. "Use of Agrochemicals in Agriculture: Alarming Issues and Solutions." *Input Use Efficiency for Food and Environmental Security*. Singapore: Springer Nature. pp. 85-122.

Nowak, B. "Precision Agriculture: Where do We Stand? A Review of the Adoption of Precision Agriculture Technologies on Field Crops Farms in Developed Countries." *Agricultural Research* 10, 4 (2021): 515-22.

Onyango, C.M., Nyaga, J.M., Wetterlind, J., Söderström, M., Piikki, K. "Precision Agriculture for Resource Use Efficiency in Smallholder Farming Systems in Sub-Saharan Africa: A Systematic Review." *Sustainability* 13, 3 (2021): 1158.

Pérez-Ruiz, M., Agüera, J., Gil, J.A., Slaughter, D.C. "Optimization of agrochemical application in olive groves based on positioning sensor." *Precision Agriculture* 12, 4 (2011): 564-75.

Rayhan Shaheb, M., Sarker, A., Shearer, S.A. 2022. "Precision Agriculture for Sustainable Soil and Crop Management." *Soil Science—Emerging Technologies, Global Perspectives and Applications*. IntechOpen. https://doi.org/10.5772/intechopen.101759 [Last accessed February, 2024].

Suruliandi, A., Mariammal, G., Raja, S.P. "Crop prediction based on soil and environmental characteristics using feature selection techniques." *Mathematical and Computer Modelling of Dynamical Systems* 27, 1 (2021): 117-40.

Tayari, E., Jamshid, A.R., Goodarzi, H.R. "Role of GPS and GIS in precision agriculture." *Journal of Scientific Research and Development* 2, 3 (2015): 157-62.

Thompson, N.M., Bir, C., Widmar D.A., Mintert J.R. "Farmer Perceptions of Precision Agriculture Technology Benefits." *Journal of Agricultural and Applied Economics* 51, 1 (2019): 142-63.

Tsuchikawa, S., Ma, T., Inagaki, T. "Application of near-infrared spectroscopy to agriculture and forestry." *Analytical Sciences* 38, 4 (2022): 635-42.

Weißhuhn, P., Helming, K., Ferretti, J.. "Research impact assessment in agriculture—A review of approaches and impact areas." *Research Evaluation* 27, 1 (2018): 36-42.

12

Nanoparticles

ABSTRACT

Nanoparticles are tiny particles that have at least one dimension between 1 and 100 nanometers. They have different physical, chemical and biological properties than larger particles. Depending on their size, nanoparticles can have various applications, such as protecting crops, emitting light or developing sensors. Nanotechnology is a promising field that has many uses in agriculture. However, the environmental impact of nanoparticles is still unclear and needs more research.

Keywords: Nanoparticles, nanotechnology, agriculture

Introduction

Nanotechnology is a scientific and technological field that manipulates matter at the nanoscale to create, design, characterize, and use structures, systems, and devices with novel properties (Sattler 2016). Nanoparticles are one of the most important products of nanotechnology, as they exhibit different features from their bulk counterparts due to their size and shape (Sabir et al., 2014). Nanoparticles have applications in various disciplines, such as materials science, medicine, agriculture, physics, and chemistry. Nanoparticles have unique electronic, optical, and chemical properties as well as mechanical characteristics, that depend on their high surface to volume ratios and their enhanced reactivity at the molecular level. Therefore, they have attracted considerable research interest for both fundamental and practical purposes. There are various methods to synthesize nanoparticles, such as mechanical, chemical, and other approaches. However, the characterization of nanoparticles is not always straightforward and reliable. This is because nanoparticles pose some challenges for the analytical techniques that are commonly used for bulk materials, such as their small size and low quantity in some cases. Moreover, not every research team has easy access to a wide range of characterization facilities due to the interdisciplinary nature of nanoscience and nanotechnology. In fact, it is often necessary to use a comprehensive approach that combines complementary techniques to characterize nanoparticles more thoroughly. It is also useful to know the strengths and limitations of each technique in order to determine if using one or two of them is sufficient to obtain accurate information for a specific parameter (e.g., particle size). The scientific community is aware that there may be some differences between the way analytical characterization methods work for nanoparticles and their more "traditional" modes of use for more "conventional" (macroscopic) materials. Nanoscience and nanotechnology are still evolving fields that require constant development of new characterization methods.

Characterization of Nanoparticles

Spectroscopic Analysis (UV-Visible Spectroscopy)

Ultraviolet (UV)/visible spectroscopy is a method that measures how much light is absorbed and scattered by a material. This quantity is called the extinction, which is the sum of the light that does not pass through the material. The basic principle of this method is to place a sample between a light source and a detector, and measure the intensity of the UV/visible light before and after it hits the sample. By comparing these data at different wavelengths, the extinction spectrum of the sample can be obtained. The extinction spectrum shows how the extinction varies with the wavelength of the light. To avoid including the effects of the solvent or other substances in the sample, a blank solution is used to correct the background. UV/visible spectroscopy is a valuable technique for detecting, characterizing, and studying nanomaterials (NMs), especially plasmonic nanoparticles made of gold or silver. These nanoparticles have optical properties that depend on their size, shape, concentration, aggregation state, and surrounding medium. These properties affect how they interact with light and produce distinctive extinction spectra. The measured spectra can be compared with theoretical models to understand the optical behavior of the nanoparticles. Nonplasmonic nanoparticles also have optical properties that depend on their size and concentration, but their spectra are less sensitive to their dispersion characteristics than plasmonic nanoparticles (Borowska and Jankowski 2023; Kumar 2013).

Transmission Electron Microscopy

Transmission electron microscopy (TEM) is a technique that uses an electron beam to probe a material. The electron beam interacts with the material and changes its intensity and phase depending on the thickness and composition of the material (Fig. 1). These changes create contrast in the image that reveals the structure and properties of the material. TEM can achieve much higher

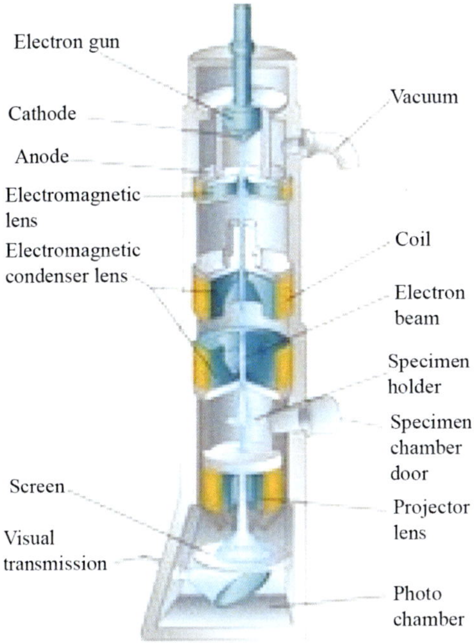

Fig. 1: Schematic image of transmission electron microscopy (TEM).

resolution than optical microscopy because electrons have much shorter wavelengths than light(Liu 2005; Ponce et al., 2012).

Dynamic Light Scattering

Dynamic light scattering (DLS) is a technique that uses a laser beam to probe the size of nanoparticles and other colloids in a solution (Fig. 2). DLS measures the fluctuations of the light intensity that is scattered by the particles as they move randomly due to Brownian motion. The speed of the particle movement depends on their size, with larger particles moving slower and scattering more light than smaller ones. By analyzing the time-dependent changes of the scattered light intensity, DLS can calculate the hydrodynamic diameter, which is the diameter of a hypothetical sphere that diffuses at the same rate as the particles under study. The hydrodynamic diameter provides information about the aggregation state of the nanoparticle solutions, which is an important parameter to complement other size measurements such as TEM. Stable and well-dispersed colloidal solutions will have hydrodynamic diameters that are similar or slightly larger than their TEM diameters, while highly aggregated

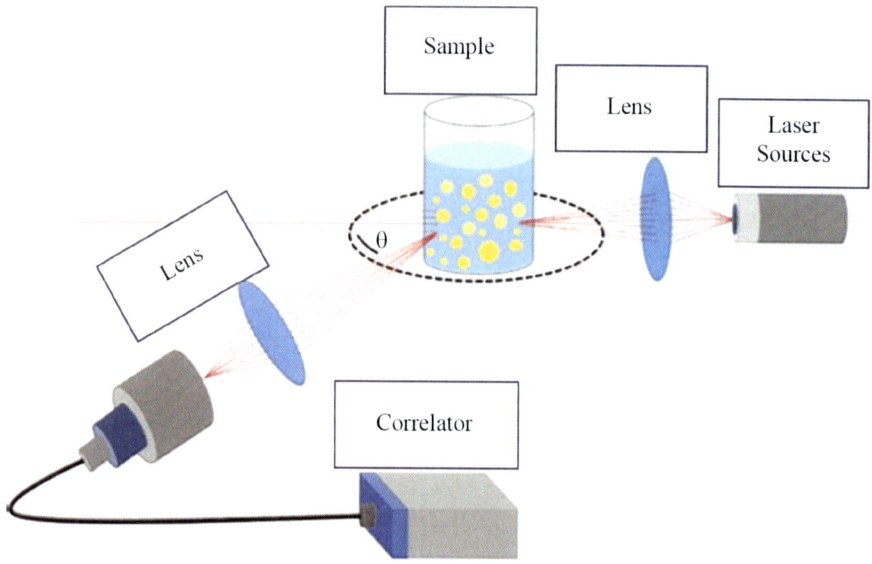

Fig. 2: Schematic image of dynamic light scattering (DLS).

solutions will have much larger hydrodynamic diameters than their TEM diameters (Carvalho et al., 2018).

Zeta Potential

Zeta potential is a measure of the effective electric charge on the surface of colloidal nanoparticles. It indicates how much the nanoparticles repel or attract each other in a solution. When the nanoparticles have a net surface charge, they attract ions of the opposite charge from the surrounding solution. These ions form a layer around the nanoparticle surface, called the electrical double layer. The zeta potential is the difference in potential between this layer and the bulk solution. The sign of the zeta potential shows whether the nanoparticles are positively or negatively charged, and the magnitude of the zeta potential shows how stable the nanoparticles are. Higher zeta potentials mean stronger electrostatic repulsion and lower tendency to aggregate (Smith et al., 2017; Vanlalveni et al., 2021). The stability of nanoparticles can be roughly classified as follows:

0–5 mV: Particles tend to agglomerate or aggregate

5–20 mV: Particles are minimally stable

20–40 mV: Particles are moderately stable

40+ mV: Particles are highly stable

Inductively Coupled Plasma Mass Spectrometry

Inductively coupled plasma mass spectrometry ICP-MS is a powerful technique for identifying and measuring the elemental composition of samples, including metals and some nonmetals with atomic masses from 7 to 250 (Clases and Gonzalez de Vega 2022). Compared to atomic absorption spectroscopy (AAS), ICP-MS offers more flexibility for simultaneous multielement analysis, higher throughput, and lower sample volume requirements (McMahon 2022). Moreover, ICP-MS can achieve detection limits for several elements in the parts per trillion (ppt) or even parts per quadrillion (ppq) range (Clases and Gonzalez de Vega 2022). However, the sensitivity of the measurement may be reduced by interference effects from co-contaminants or plasma interactions. Many of these interferences can be eliminated by using ICP-MS instruments equipped with a collision cell, such as the one used at NanoComposix (Engelhard 2011) To ionize elements by stripping electrons from individual atoms, liquid samples are aerosolized in a nebulizer and introduced into a high temperature heat source (such as argon plasma). The ionized atoms are then separated from neutral particles in a vacuum chamber by an ion guide and detected by a mass spectrometer (MS). The elements are discriminated by their mass-to-charge ratio, and their concentration is calculated by comparing the intensity of each element's ion signal to an internal calibration standard (Bolea-Fernandez et al., 2017). Sample digestion may be necessary because ICP-MS samples have to be in a liquid form to enable ionization. The amount of the element of interest that needs to be present in the sample depends on its concentration, any potential interferences, the instrument's sensitivity to the element, the number of analyses being performed, and the total number of elements of interest. Sample volumes as low as 100 uL can be analyzed.

Dark Field Microscopy

Dark field microscopy is a light microscopy technique that uses oblique incident illumination to image specimens. The background of the image is black because the objective lens does not collect the incoming beam except where it has been scattered by the specimen. This method is widely used for studying plasmonic nanoparticles, which have very weak scattered light compared to the incident beam (Gao et al., 2021). The following are dark field images of 60 nm silver nanoparticles and silver nanowires (Shang et al., 2022). Because the diameter of a 60 nm silver nanoparticle is smaller than the wavelength of light and light microscopy is limited by diffraction, the nanoparticles appear as bright point sources of light (Wang et al., 2019).

Aerodynamic Particle Sizer

Aerodynamic particle size (APS) analysis is a method for measuring the size of particles or droplets in the air that have diameters between 500 nm and 20 μm (Yoshida et al., 2020). APS analysis is based on the principle that the terminal velocity of a particle depends on its size and shape. The APS is defined as the diameter of a spherical particle with the same density and terminal velocity as the actual particle (Roberts and Mitchell 2019). The terminal velocity can be measured by timing how long it takes a particle to travel between two laser beams at a known distance apart (Mitchell and Nagel 1999). The samples have to be aerosolized and introduced into a chamber where the size distribution and optical properties of the particles can be determined (Pfeifer et al., 2016).

Scanning Mobility Particle Sizer

Aerodynamic particle size analysis measures the change in the terminal velocity of aerosolized particles or droplets. The terminal velocity depends on the size and shape of the particles or droplets. The APS is the diameter of a spherical particle with the same density and terminal velocity as the actual particle (Hao et al., 2023). The terminal velocity can be measured by timing how long it takes a particle to travel between two laser beams at a known distance apart (Mei et al., 2021). The samples have to be aerosolized and introduced into a chamber where the size distribution and optical properties of the particles can be determined (Sipilä et al., 2008). The particles are then individually counted after condensation (Weber et al., 2023). A distribution of particle sizes between 10 and 500 nm is obtained by measuring the number of particles at each voltage (Mei et al., 2021).

Matrix-assisted Laser Desorption/Ionization Mass Spectrometry

Matrix-assisted laser desorption ionization time-of-flight mass spectrometry (MALDI-tof-MS) is a technique that uses a laser to ionize molecules from a solid or liquid matrix and measure their mass-to-charge ratio by a time-of-flight detector (Do et al., 2022). This technique allows the analysis of small to large organic molecules and biomolecules with minimal fragmentation, enabling the accurate determination of their molecular weight from simple to complex samples (Gómez-Velásquez et al., 2021). MALDI-tof-MS can be applied to various fields, such as proteomics, peptide mass fingerprinting, polymer analysis, and tissue imaging (Elbehiry et al.,

2022). For low-molecular-weight peptide or chemical samples, the conventional MALDI analysis method uses a 2,5-dihydroxybenzoic acid (DHB) matrix; for medium-to-high-molecular-weight proteins or peptides, it uses a sinapic acid (SA) or α-cyano-4-hydroxycinnamic acid (CHCA) matrix (Feucherolles et al., 2019). NanoComposix is developing novel nanoparticle-based matrices that have advantages over organic matrices. Gold nanoparticles can often detect chemical signatures that are difficult to analyze with conventional organic matrices due to their high sensitivity to classes of molecules such as lipids. Moreover, nanoparticle matrices require only 1/100th the mass of an organic matrix. NanoComposix is conducting research on tissue imaging using both MALDI matrices and nanoparticles.

Synthesis of Nanoparticles

The "bottom-up" strategy and the "top-down" approach are two alternate methods for creating metallic nanoparticles.

Bottom-up Strategy

The bottom-up approach uses chemical or biological methods to create nanoparticles of different materials and then combines them into nanostructured materials (Baig et al., 2021). One of the most common and oldest methods for producing metallic nanoparticles is wet chemical synthesis, which involves reducing metal ions in a liquid medium containing various reactants (Indiarto et al., 2022). A stabilizing agent, such as sodium dodecyl benzyl sulfate or polyvinyl pyrrolidone, is also added to the reaction mixture to prevent the metallic nanoparticles from clumping together. However, the chemical methods have both advantages and disadvantages, such as high yield, low cost, easy scalability, but also low purity, high toxicity, and limited control over size and shape (Escudero et al., 2021).

Top-down Approach

The top-down approach is a method of synthesizing nanostructures by breaking down a larger material into smaller ones. For the synthesis of metallic nanoparticles, various physical or chemical methods can be used, such as attrition and pyrolysis (Baig et al., 2021). Attrition is a process of grinding macro- or microscale particles using a device such as a ball mill. The resulting particles are then air-classified to obtain nonoxidized nanoparticles. The factors that affect the properties of the nanoparticles include the type and duration of milling, and the atmospheric media. Pyrolysis is a process of burning an organic precursor that is forced through an orifice under high pressure (de Oliveira et al., 2020). The resulting ash is air-classified to obtain oxidized nanoparticles.

Applications of Nanoparticles

Crop Protection with Nanotechnology

The world is facing a major challenge of feeding a growing population that is expected to reach 9 billion by 2050, while coping with the impacts of climate change, urbanization, and environmental degradation (Nile et al., 2020). The natural resources that are essential for food production, such as land, water, and soil, are becoming scarce and degraded. The cost of chemical fertilizers, pesticides, and other inputs is also increasing due to the depletion of fossil fuels (Mittal et al., 2020). To overcome these limitations, there is a need for innovative technologies that can improve agricultural productivity, efficiency, and sustainability. Nanotechnology is one of the promising technologies that can offer solutions for various aspects of crop production, such as nutrient management, pest control, water conservation, and quality enhancement (Kumar et al., 2019). Nanotechnology involves the manipulation of matter at the nanoscale (1–100 nm), which results in novel properties and functions that can be exploited for various applications (Huang et al., 2015). Nanotechnology can also enable precision farming techniques, which involve the use of sensors, drones, robots, and smart systems to monitor and optimize crop growth and health (Liu et al., 2021).

Enhancing Plant Development with Nanoparticles

Plant germination and growth can benefit greatly from the usage of NM. The carbon nanotubes have the potential to control plant development and seed germination. At a range of concentrations from 5 to 500 g/mi, multiwalled carbon nanotubes (MWCNTs) have the ability to increase the development of tobacco cell culture by 55%–64% as compared to control. While activated carbon significantly hindered cellular growth at higher concentrations, it stimulated cell proliferation at lower one.

Pathogen Control with Nanotechnology

Nanomaterials are materials that have at least one dimension in the nanoscale (1–100 nm). NM can have different properties and functions than their bulk counterparts, which can be useful for various applications. One of the applications of NM is to enhance plant germination and growth by affecting their physiological and biochemical processes. Carbon nanotubes (CNTs) are one type of NMs that have been studied for their effects on plants. CNTs are cylindrical structures made of carbon atoms that have high aspect ratios, electrical conductivity, mechanical strength, and thermal stability. CNTs can interact with plant cells and tissues in different ways, such as adsorption, uptake, translocation, accumulation, and degradation. CNTs can also influence the expression of genes that are involved in cell division and plant development (Mathew et al., 2021). One of the studies that investigated the effects of CNTs on plants was conducted by Khodakovskaya et al., (Khodakovskaya et al., 2012). They exposed tobacco cell cultures to different concentrations of MWCNTs, which are CNTs that consist of multiple layers of CNTs. They found that MWCNTs increased the growth of tobacco cells by 55–64% compared to control. They also observed that MWCNTs enhanced the activity of some enzymes that are related to plant metabolism and stress response. However, they noted that activated carbon, which is a material similar to CNTs but without nanoscale structure, had opposite effects on tobacco cells. Activated carbon inhibited cell growth at higher concentrations, but stimulated cell proliferation at lower concentrations. They suggested that the different effects of MWCNTs and activated carbon were due to their different interactions with plant cell walls and membranes.

Using Nanotechnology for Pest Control

Nanopesticide formulations are designed to improve the delivery and performance of active ingredients that have low solubility or stability in water. By using additives or nanoscale structures, nanopesticide formulations can increase the bioavailability of poorly water-soluble agrochemicals, which means that they can be more effective and require lower doses. One example of a poorly water-soluble agrochemical is rotenone, which is a natural insecticide that has been used for decades to control aphids and thrips (Xu et al., 2022). However, rotenone has limited applicability because it is unstable, degrades, and isomerizes when exposed to sunlight (Deka et al., 2021). To overcome these drawbacks, researchers have developed novel amphiphilic chitosan derivatives that can form nanoparticles with rotenone. Amphiphilic means that the molecules have both hydrophilic and hydrophobic parts, which can help them dissolve in both water and oil. Chitosan is a natural polymer derived from chitin, which is found in the shells of crustaceans and insects. Chitosan has many advantages as a nanocarrier, such as biodegradability, biocompatibility, antimicrobial activity, and low toxicity (Chaud et al., 2021). One of the novel amphiphilic chitosan derivatives is N-(octadecanol-1-glycidyl ether)-O sulfate chitosan (NOSCS), which was synthesized by attaching octadecanol-1-glycidyl ether to amino groups and sulphate to hydroxyl groups of chitosan. NOSCS can form stable nanoparticles with rotenone that have improved solubility, stability, and insecticidal activity compared to pure rotenone (Pan et al., 2023).

Conclusion

Nanotechnology can regulate the delivery of agrochemicals to enhance plant growth, nutrient uptake, and disease resistance. Nano encapsulation improves the effectiveness, selectivity, and sustainability of pesticides, herbicides, and insecticides. Nanotechnology also increases the solubility of the active ingredients, which boosts their activity in the targeted area. By combining nanotechnology with green chemistry and biotechnology, NMs can be manufactured using plant extracts and live cells, which reduces the use of hazardous solvents and creates a more sterile environment. Nanotechnology has already been widely used in cosmetic and pharmaceutical products, but its toxicity is still not fully understood. Therefore, more research is needed before introducing NMs for crop protection. Nanotechnology offers a promising opportunity to develop innovative and eco-friendly solutions for agriculture.

References

Baig, N., Kammakakam, I., Falath, W. "Nanomaterials: a review of synthesis methods, properties, recent progress, and challenges." *Materials Advances* 2, 6 (2021):

Bolea-Fernandez, E., Balcaen, L., Resano, M., Vanhaecke, F. "Overcoming spectral overlap via inductively coupled plasma-tandem mass spectrometry (ICP-MS/MS). A tutorial review." *Journal of Analytical Atomic Spectrometry* 32, 9 (2017): 1660-79.

Borowska, M., Jankowski, K. "Basic and advanced spectrometric methods for complete nanoparticles characterization in bio/eco systems: current status and future prospects." *Analytical and Bioanalytical Chemistry* 415, 18 (2023): 4023-38.

Carvalho, P.M., Felício, M.R., Santos, N.C., Gonçalves, S., Domingues, M.M. "Application of Light Scattering Techniques to Nanoparticle Characterization and Development." *Frontiers in Chemistry* 6 (2018).

Chaud, M., Souto, E.B., Zielinska, A., Severino, P., Batain, F., Oliveira-Junior, J., et al. "Nanopesticides in Agriculture: Benefits and Challenge in Agricultural Productivity, Toxicological Risks to Human Health and Environment." *Toxics* 9, 6 (2021): 131.

Clases, D., Gonzalez de Vega, R. "Facets of ICP-MS and their potential in the medical sciences—Part 1: fundamentals, stand-alone and hyphenated techniques." *Analytical and Bioanalytical Chemistry* 414, 25 (2022): 7337-61.

de Oliveira, P.F.M., Torresi, R.M., Emmerling, F., Camargo, P.H.C. "Challenges and opportunities in the bottom-up mechanochemical synthesis of noble metal nanoparticles." *Journal of Materials Chemistry A* 8, 32 (2020): 16114-41.

Deka, B., Babu, A., Baruah, C., Barthakur, M. "Nanopesticides: A Systematic Review of Their Prospects With Special Reference to Tea Pest Management." *Frontiers in Nutrition* 8 (2021).

Do, T., Guran, R., Adam, V., Zitka, O. "Use of MALDI-TOF mass spectrometry for virus identification: a review." *The Analyst* 147, 14 (2022): 3131-54.

Elbehiry, A., Aldubaib, M., Abalkhail, A., Marzouk, E., ALbeloushi, A., Moussa, I., et al. "How MALDI-TOF Mass Spectrometry Technology Contributes to Microbial Infection Control in Healthcare Settings." *Vaccines* 10, 11 (2022): 1881.

Engelhard, C. "Inductively coupled plasma mass spectrometry: recent trends and developments." *Analytical and Bioanalytical Chemistry* 399, 1 (2011): 213-9.

Escudero, A., Carrillo-Carrión, C., Romero-Ben, E., Franco, A., Rosales-Barrios, C., Castillejos, M. C., et al. "Molecular Bottom-Up Approaches for the Synthesis of Inorganic and Hybrid Nanostructures." *Inorganics* 9, 7 (2021): 58.

Feucherolles, M., Poppert, S., Utzinger, J., Becker, S.L. "MALDI-TOF mass spectrometry as a diagnostic tool in human and veterinary helminthology: a systematic review." Parasites & Vectors 12, 1 (2019): 245.

Gao, P.F., Lei, G., Huang, C.Z. "Dark-Field Microscopy: Recent Advances in Accurate Analysis and Emerging Applications." *Analytical Chemistry* 93, 11 (2021): 4707-26.

Gómez-Velásquez, J.C., Mojica-Figueroa, I.L., Santos, C., Lima, N., Mesa-Arango, A.C. "MALDI-TOF MS: Foundations and a Practical Approach to the Clinically Relevant Filamentous Fungi Identification." *Current Fungal Infection Reports* 15, 4 (2021): 162-70.

Hao, W., Mei, F., Hering, S., Spielman, S., Schmid, B., Tomlinson, J., Wang, Y. "Mapping the performance of a versatile water-based condensation particle counter (vWCPC) with numerical simulation and experimental study." *Atmospheric Measurement Techniques* 16, 17 (2023): 3973-86.

Huang, S., Wang, L., Liu, L., Hou, Y., Li, L. "Nanotechnology in agriculture, livestock, and aquaculture in China. A review." *Agronomy for Sustainable Development* 35, 2 (2015): 369-400.

Indiarto, R., Indriana, L.P.A., Andoyo, R., Subroto, E., Nurhadi, B. "Bottom–up nanoparticle synthesis: a review of techniques, polyphenol-based core materials, and their properties." *European Food Research and Technology* 248, 1 (2022): 1-24.

Khodakovskaya, M.V., de Silva, K., Biris, A.S., Dervishi, E., Villagarcia, H. "Carbon Nanotubes Induce Growth Enhancement of Tobacco Cells." *ACS Nano* 6, 3 (2012): 2128-35.

Kumar, A., Gupta, K., Dixit, S., Mishra, K., Srivastava, S. "A review on positive and negative impacts of nanotechnology in agriculture." *International Journal of Environmental Science and Technology* 16, 4 (2019): 2175-84.

Kumar, C. 2013. *UV-VIS and Photoluminescence Spectroscopy for Nanomaterials Characterization*. Germany: Springer Berlin Heidelberg.

Liu, C., Zhou, H., Zhou, J. "The Applications of Nanotechnology in Crop Production." *Molecules* 26, 23 (2021): 7070.

Liu, J. "Scanning transmission electron microscopy and its application to the study of nanoparticles and nanoparticle systems." *Microscopy* 54, 3 (2005): 251-78.

Mathew, S., Tiwari, D.K., Tripathi, D. "Interaction of carbon nanotubes with plant system: a review." *Carbon Letters* 31, 2 (2021): 167-76.

McMahon, G. 2024. ICP-MS—ICP-MS Instrumentation, ICP-MS Analysis, Strengths and Limitations. Technology Networks. https://www.technologynetworks.com/analysis/articles/ICP-MS-ICP-MS-Instrumentation-ICP-MS-Analysis-Strengths-and-Limitations-341882 [Last accessed February, 2024].

Mei, F., Spielman, S., Hering, S., Wang, J., Pekour, M.S., Lewis, G., et al. "Simulation-aided characterization of a versatile water-based condensation particle counter for atmospheric airborne research." *Atmospheric Measurement Techniques* 14, 11 (2021): 7329-40.

Mitchell, J.P., Nagel, M.W. "Time-of-Flight Aerodynamic Particle Size Analyzers: Their Use and Limitations for the Evaluation of Medical Aerosols." *Journal of Aerosol Medicine* 12, 4 (1999): 217-40.

Mittal, D., Kaur, G., Singh, P., Yadav, K., Ali, S.A. "Nanoparticle-Based Sustainable Agriculture and Food Science: Recent Advances and Future Outlook." *Frontiers in Nanotechnology* 2 (2020).

Nile, S.H., Baskar, V., Selvaraj, D., Nile, A., Xiao, J., Kai, G. "Nanotechnologies in Food Science: Applications, Recent Trends, and Future Perspectives." *Nano-Micro Letters*, 12, 1 (2020):e 45.

Pan, X., Guo, X., Zhai, T., Zhang, D., Rao, W., Cao, F., et al. "Nanobiopesticides in sustainable agriculture: developments, challenges, and perspectives." *Environmental Science: Nano* 10, 1 (2023): 41-61.

Pfeifer, S., Müller, T., Weinhold, K., Zikova, N., Martins dos Santos, S., Marinoni, A., et al. Intercomparison of 15 aerodynamic particle size spectrometers (APS 3321): uncertainties in particle sizing and number size distribution. *Atmospheric Measurement Techniques* 9, 4 (2016): 1545-51.

Ponce, A., Mejía-Rosales, S., José-Yacamán, M. "Scanning Transmission Electron Microscopy Methods for the Analysis of Nanoparticles." *Methods in Molecular Biology* 906 (2012): 453-71.

Roberts, D.L., Mitchell, J.P. "Measurement of Aerodynamic Particle Size Distribution of Orally Inhaled Products by Cascade Impactor: How to Let the Product Specification Drive the Quality Requirements of the Cascade Impactor." *AAPS PharmSciTech* 20, 2 (2019): 57.

Sabir, S., Arshad, M., Chaudhari, S.K. "Zinc oxide nanoparticles for revolutionizing agriculture: Synthesis and applications." *Scientific World Journal* 2014.

Sattler, K.D. 2010. *Handbook of Nanophysics*. Boca Raton: CRC Press.

Shang, J., Fan, J., Qin, W., Li, K. "Single-Particle Measurements of Nanocatalysis with Dark-Field Microscopy." *Catalysts* 12, 7 (2022).

Sipilä, M., Lehtipalo, K., Kulmala, M., Petäjä, T., Junninen, H., Aalto, P.P., et al. "Applicability of condensation particle counters to measure atmospheric clusters." *Atmospheric Chemistry and Physics* 8, 14 (2008): 4049-60.

Smith, M.C., Crist, R.M., Clogston, J.D., McNeil, S.E. "Zeta potential: a case study of cationic, anionic, and neutral liposomes." *Analytical and Bioanalytical Chemistry*. 409, 24 (2017): 5779-87.

Vanlalveni, C., Lallianrawna, S., Biswas, A., Selvaraj, M., Changmai, B., Rokhum, S.L. "Green synthesis of silver nanoparticles using plant extracts and their antimicrobial activities: a review of recent literature." *RSC Advances* 11, 5 (2021): 2804-37.

Wang, B., Sun, D., Zhang, C., Wang, K., Bai, J. "Dark-field microscopy for characterization of single molecule dynamics in vitro and in vivo." *Analytical Methods* 11, 21 (2019) 2778-84.

Weber, P., Bischof, O.F., Fischer, B., Berg, M., Schmitt, J., Steiner, G., et al. "A new working fluid for condensation particle counters for use in sensitive working environments." *Aerosol Research* 1, 1 (2023): 1-12.

Xu, Z., Tang, T., Lin, Q., Yu, J., Zhang, C., Zhao, X., et al. "Environmental risks and the potential benefits of nanopesticides: a review." *Environmental Chemistry Letters* 20, 3 (2022): 2097-108.

Yoshida, H., Usui, A., Abe, Y., Goda, Y., Izutsu, K. "Relationship Between Geometric and Aerodynamic Particle Size Distributions in the Formulation of Solution and Suspension Metered-Dose Inhalers." *AAPS PharmSciTech* 21, 5 (2020): 158.

13

Nanofertilizers

Abstract

One of the most significant innovations in nanotechnology is nanofertilizers, which are designed to be an alternative to traditional fertilizers and a major step toward precision agriculture. Nanofertilizers have the following features:

- They release nutrients slowly and steadily, unlike conventional fertilizers that release nutrients quickly and unevenly
- They have a size range of 0.1–100 nm, which is similar to the size of plant cell wall (20 nm), making them easier to be absorbed by plants
- They aim to achieve cost efficiency, sustainability, eco-friendliness, superiority over conventional fertilizers, minimal wastage, and almost no pollution

Nanofertilizers consist of nanomaterials that contain the nutrients to be supplied. They can be classified into three main types based on their composition and function:

- *Nanoscale fertilizers*: These are pure nanomaterials that provide nutrients directly to plants
- *Nanoscale additive fertilizers*: These are nanomaterials that are added to conventional fertilizers to enhance their performance and efficiency
- *Nanoscale coating fertilizers*: These are nanomaterials that are coated on conventional fertilizers to control their release and reduce their leaching

Nanofertilizers can also be classified based on the type of nutrients they provide:

- *Micronutrient nanofertilizers*: These provide essential trace elements such as iron, zinc, copper, etc.
- *Macronutrient nanofertilizers*: These provide major elements such as nitrogen, phosphorus, and potassium
- *Bionanofertilizers*: These provide beneficial microorganisms, such as bacteria and fungi

The main reason for developing nanofertilizers is to overcome the problems caused by conventional fertilizers, such as pollution, water, and soil degradation, excessive use leading to nutrient imbalance, toxicity, and organic matter depletion. Nanofertilizers have been shown to improve nutrient use efficiency (NUE), reduce pollution, enhance crop yield and quality, etc. However, nanofertilizers also have some drawbacks that need to be addressed. One of the main challenges is the safety of nanomaterials for human health and the environment. There is still a lack of sufficient research on the potential risks and impacts of nanomaterials on crops that are consumed by humans. Another challenge is the high cost of nanofertilizers, which makes them inaccessible for many farmers. Therefore, more research and development is needed to make nanofertilizers more safe, affordable and widely available. Nanofertilizers are a promising option for sustainable agriculture, but they still require further improvement and evaluation.

Keywords: Nanofertilizers, nanomaterials, top-down approach, bottom-up approach, biosynthesis, NUE

Introduction

Nanofertilizers are products of nanotechnology that are created or modified from conventional fertilizers, bulk fertilizer materials, or plant extracts. They use different methods such as chemical, physical, mechanical, or biological processes to produce nanosized particles that can enhance soil fertility, crop productivity, and quality (KhetiGaadi 2022). Nanofertilizers have a size range of 0.1–100 nm, which is similar to the size of plant cell wall (20 nm), making them easier to be absorbed by plants (Tiwari 2022). As shown in Figure 1, nanofertilizers can also improve NUE, reduce environmental pollution, and lower the cost of fertilization. Nanofertilizers are a new and improved alternative to conventional fertilizers that provide nutrients in optimal amounts and in much smaller quantities. Nanofertilizers use nanoparticles, which are extremely tiny particles with a size of one billionth of a meter or 0.1–100 nm. Nanoparticles have a similar size to the plant cell wall (20 nm), which makes them more easily absorbed by plants. Nanofertilizers are products of nanotechnology, which is the science and engineering of manipulating matter at the nanoscale. Nanotechnology can enhance the nutrient efficiency and plant nutrition of nanofertilizers compared to traditional fertilizers.

Nanofertilizers have become the new and better replacement for the conventional fertilizers providing the nutrients in the required amounts, in a lot lesser quantities as compared to conventional ones. Nanofertilizers are also based on the principle of nanoparticles. The word “nano” means billionth, so

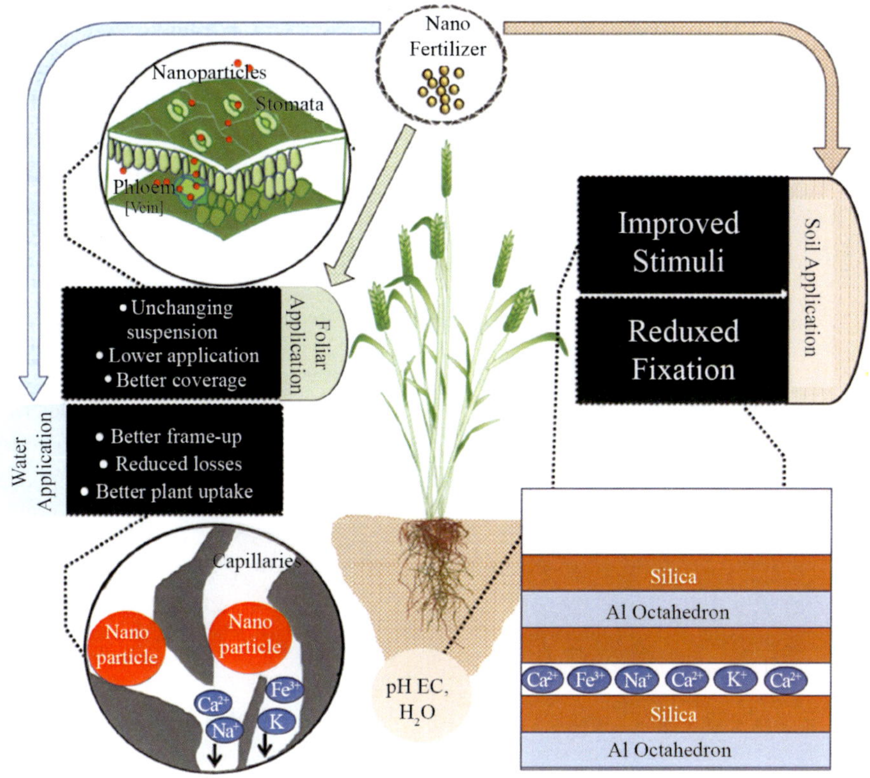

Fig. 1: Merits of nanofertilizers.

the term nanoparticles refers to particles belonging in the size range of billionth of a meter. Typically, the size of nanoparticles belongs in the range of 0.1–100 nm. A nanofertilizer is any product that is made with nanoparticles or uses nanotechnology to improve nutrient efficiency. These are being widely studied to increase the nutrient efficiency and enhance plant nutrition compared to traditional fertilizers (Rajput et al., 2021). Nanofertilizers have a slow-release nutrient effect, which means they provide nutrients to plants gradually and steadily. They are usually coated with some materials that help them stick to the soil better. Nanofertilizers also have a high surface area to volume ratio, which is a common feature of nanoparticles. This means they have more contact with the soil and plants and can deliver nutrients more efficiently (Production et al., 2017).

Nanofertilizers use different delivery systems to provide nutrients to plants in a more efficient and controlled way. Some of the delivery systems are:

- *Chemical system*: This involves coating the nanofertilizers with chemicals such as sulfur to slow down their release. For example, urea and phosphorus fertilizers can be coated with nanomaterials to improve their performance and stability
- *Biological system*: This involves using biological agents such as neem oil to coat the nanofertilizers. The biological system also includes the biosynthesis of nanofertilizers using microorganisms or plants
- *Field application of fertilizers*: This involves applying the nanofertilizers in the best way to maximize their effect. For instance, the problem of droplet size in foliar sprays can be solved by using nanosized fertilizers that can penetrate the plant leaves better
- *Controlled release of nanofertilizers*: This involves combining the nanofertilizers with other materials such as hydrogels, special films, or biopolymers that can regulate their release. Encapsulation is another technique that can control the release of nanofertilizers by enclosing them in a protective layer

Nanofertilizers are a better and more beneficial option than traditional fertilizers, and they are needed to address the various challenges faced by the agricultural sector. Some of the challenges that affect crop production and quality are:

- Low crop yields due to imbalanced fertilization and soil degradation
- Reduced soil organic matter content due to excessive use of chemical fertilizers and erosion
- High levels of nutrient deficiencies in crops due to poor nutrient availability and uptake
- Adverse climate change effects such as drought, heat, salinity, pests and diseases
- Limited availability of arable land and water resources due to population growth and urbanization
- Resistance to genetically modified organisms (GMOs) due to ethical, environmental, and health concerns
- Shortage of labor due to migration, aging, and mechanization

Nanofertilizers can help overcome these challenges by providing nutrients in optimal amounts and in smaller quantities, enhancing soil fertility and crop productivity, reducing environmental pollution and cost of fertilization, and improving crop quality and resilience. Nanofertilizers can also improve NUE, reduce nutrient losses, and increase nutrient mobility in the soil (Mohan et al., 2022). Some of the advantages of using nanofertilizers instead of conventional ones are:

- Nanofertilizers provide nutrients to plants in a controlled and gradual way, unlike chemical fertilizers that release nutrients quickly and unevenly
- Nanofertilizers have a higher nutrient efficiency and utilization, as they have less losses due to leaching and volatilization
- Nanofertilizers have a better uptake by plants, as they can pass through the nanosized pores and the molecular transporters of the plant cells. Nanofertilizers can also use different ion channels to increase nutrient absorption. Moreover, nanofertilizers can move through the plasmodesmata, which are the connections between plant cells, and deliver nutrients to the places where they are needed
- Nanofertilizers can be applied in smaller amounts than synthetic fertilizers, as they have less wastage and more effectiveness
- Nanofertilizers have a lower environmental impact, as they have less pollution and contamination due to their small losses
- Nanofertilizers have a higher solubility and diffusion, which makes them superior to conventional synthetic fertilizers
- Nanofertilizers can be smart, such as polymer-coated fertilizers, which prevent the premature contact of nutrients with soil and water. These fertilizers release nutrients only when the plants are ready to take them up

Compositions

Nanofertilizers are composed of nanoparticles that contain nutrients for plants. These nanoparticles are produced by either physical methods, such as breaking down larger particles into smaller ones, or chemical methods, such as synthesizing new particles from smaller molecules. The components of nanofertilizers depend on the type and need of the fertilizer and the crop that it is applied to. For example, nanosilicon dioxide is used for maize, nanozinc oxide is used for chickpea, nanoiron is used for cotton, and nanoiron slag powder is used for maize to improve its quality (Butt and Naseer 2020). Nanofertilizers can enhance the absorption and utilization of nutrients by plants, as they have a similar size to the plant cell wall (20 nm), which makes them more easily taken up by plants (Fatima et al., 2021). Nanofertilizers can also release nutrients in a controlled and gradual way, unlike conventional fertilizers that release nutrients quickly and unevenly (Fatima et al., 2021; Iqbal et al., 2019). Nanofertilizers can be made by encapsulating or coating nutrients with nanomaterials, such as polymers, clays, metals, or carbon-based materials (Fatima et al., 2021; Pati et al., 2021). There are three main ways of encapsulating or coating nutrients with nanomaterials (Nongbet et al., 2022):

- Plant nutrients can be enclosed within nanomaterials of different types and compositions
- Nutrient particles can be covered with a thin layer of nanomaterials, such as polymer films
- Nutrients can also be delivered in the form of emulsions or particles that have a size range of nanoparticles

Types

Nanofertilizers can be classified into three main types based on their composition and function:

1. *Nanoscale fertilizers*: These are conventional fertilizers that are reduced in size to the nanoscale. They have the same chemical composition as the original fertilizers, but they have a higher surface area to volume ratio, which makes them more soluble and diffusible
2. *Nanoscale additive fertilizers*: These are conventional fertilizers that contain some additional nanomaterials, such as metals, clays, polymers, or carbon-based materials. The nanomaterials can enhance the performance and efficiency of the fertilizers by increasing their stability, mobility, or availability
3. *Nanoscale coating fertilizers*: These are nutrients that are encapsulated or coated by nanomaterials, such as polymer films or nanoscale pores. The nanomaterials can control the release and delivery of the nutrients by preventing their premature contact with soil and water

Nanofertilizers can also be classified based on the type of nutrients they provide :

- *Macronutrient nanofertilizers*: These provide major elements such as nitrogen (N), phosphorus (P), potassium (K), magnesium (Mg), sulfur (S) and calcium (Ca) in encapsulated form with specific nanomaterials. These elements are essential for plant growth and development, but they are often deficient or imbalanced in the soil

- *Micronutrient nanofertilizers*: These provide essential trace elements, such as iron, manganese, zinc, copper, in encapsulated form with specific nanomaterials. These elements are required in small amounts by plants, but they can have a significant impact on plant health and quality
- *Nanobiofertilizers*: These provide beneficial microorganisms, such as bacteria and fungi, in combination with nanostructures or nanoparticles. These microorganisms can improve soil fertility and plant nutrition by fixing atmospheric nitrogen, solubilizing phosphorus, or producing growth-promoting substances. Nanotechnology can enhance the viability, stability, and delivery of these microorganisms

Nanofertilizers are a promising option for sustainable agriculture, as they can improve NUE, reduce environmental pollution, and lower the cost of fertilization. However, nanofertilizers also have some challenges and risks that need to be addressed, such as the safety of nanomaterials for human health and the environment, the regulation and standardization of nanofertilizers, and the accessibility and affordability of nanofertilizers for farmers. Therefore, more research and development is needed to make nanofertilizers more safe, reliable, and widely available (Pramanik et al., 2020).

Preparation of Nanofertilizers

There are different methods to synthesize nanoparticles, which are the main components of nanofertilizers (Kumar et al., 2020). These methods can be classified into three categories: top-down, bottom-up, and biosynthesis.

- *Top-down method*: This method involves breaking down bulk materials into smaller and more organized nanoparticles. It is a physical method that uses milling or grinding techniques. The drawback of this method is that it has low control over the size of nanoparticles and it produces more impurities. Some examples of top-down methods are condensation, microwave-assisted, and laser ablation methods
- *Bottom-up method*: This method involves building up nanoparticles from smaller atoms or molecules using chemical reactions. It is a chemical method that uses synthesis or precipitation techniques. The advantage of this method is that it has better control over the size of nanoparticles and it reduces impurities. Some examples of bottom-up methods are reduction, sol–gel, co-precipitation, and reverse micelle methods

- *Biosynthesis method*: This method involves using biological sources such as plants, fungi, or bacteria to produce nanoparticles. It is a biological method that uses natural processes or enzymes. The benefit of this method is that it has greater control over the toxicity and size of nanoparticles. Some examples of biosynthesis methods are plant extract, fungal extract, and bacterial extract methods

Nanoparticles can be loaded with nutrients to deliver them to specific target sites in plants (Verma, Song, Joshi, Rajput, et al., 2022; Verma, Song, Joshi, Tian, et al., 2022). Nutrient-loading on nanoparticles can be done by different ways, such as:

- *Attachment*: This involves attaching nutrients on the surface of nanoparticles using chemical bonds or physical forces
- *Adsorption*: This involves adsorbing nutrients on the surface of nanoparticles using ligands or molecules that can bind to both nutrients and nanoparticles
- *Encapsulation*: This involves enclosing nutrients within a shell or layer of nanomaterials, such as polymers, metals, or carbon-based materials
- *Entrapment*: This involves trapping nutrients within a matrix or network of nanomaterials, such as hydrogels, clays, or nanotubes
- *Synthesis*: This involves creating nutrient nanoparticles by using chemical reactions or biological processes

Potential Applications and Challenges of Nanofertilizers in Indian Agriculture

Some of the challenges and limitations of using nanofertilizers are:

- Producing nanomaterials in large quantities with consistent quality and reasonable cost; nanomaterials are often difficult and expensive to synthesize, purify, and characterize, and they may have variable properties depending on their size, shape, and surface chemistry (Butt and Naseer 2020; Fatima et al., 2021)
- Delivering nanofertilizers in a suitable form that can be integrated into the existing agricultural practices; nanofertilizers need to have the appropriate particle size, surface chemistry, compatibility, and stability to be applied to the soil or plants without causing aggregation, degradation, or toxicity (Fatima et al., 2021; Suppan 2017)
- Bridging the gap between research and application of nanofertilizers; there is a lack of sufficient research on the efficacy, safety, and

environmental impact of nanofertilizers under field conditions and across different crops and regions (Butt and Naseer 2020; Fatima et al., 2021). There is also a need for more collaboration and communication between researchers, farmers, industry, and policy makers to facilitate the development and adoption of nanofertilizers (Butt and Naseer 2020)

- Addressing the disposal and safety concerns of nanofertilizers; nanofertilizers may pose potential risks to human health and the environment due to their novel properties and interactions with biological systems (Butt and Naseer, 2020; Fatima et al., 2021). Nanofertilizers may accumulate in plant parts, soil, water, or food chains, and cause adverse effects such as oxidative stress, inflammation, or toxicity. There is a need for more assessment and monitoring of the fate and effects of nanofertilizers in the environment (Butt and Naseer 2020; Fatima et al., 2021)
- Developing and implementing appropriate regulations for nanofertilizers; nanofertilizers are not well-regulated by the existing laws and standards for conventional fertilizers or nanomaterials (Butt and Naseer 2020). There is a lack of clear definitions, classifications, guidelines, and protocols for nanofertilizers. There is also a need for more coordination and harmonization among different agencies and stakeholders involved in the regulation of nanofertilizers
- Enhancing the technical knowledge and awareness of nanofertilizers; nanofertilizers are still a new and emerging technology that requires more education and training for farmers, consumers, and regulators (Butt and Naseer 2020). Nanofertilizers are also associated with high cost and risk factors that may deter their adoption by farmers. There is a need for more dissemination and demonstration of the benefits and challenges of nanofertilizers

Nanofertilizers are fertilizers that use nanoparticles or nanotechnology to improve nutrient efficiency and plant nutrition (Avila-Quezada et al., 2022; Bhardwaj et al., 2022; Bose 2020; Sambangi et al., 2022). Nanofertilizers have several advantages over conventional fertilizers, such as:

- They are user-friendly and eco-friendly. They pose minimal risk of environmental damage by reducing the use of chemicals
- They have a positive impact on agricultural practices. They can enhance crop yield, quality, and resilience
- They can optimize the nutrient supply and delivery. They can provide nutrients in optimal amounts and in smaller quantities, reducing the wastage and losses of nutrients

- They can increase the solubility and dispersion of insoluble nutrients in soil, reduce the soil absorption and fixation, and increase the bioavailability
- They can control the release rate and pattern of nutrients for water-soluble fertilizers, avoiding the premature or uneven release of nutrients
- They can reduce the loss rate of fertilizer into soil by leaching, volatilization, or runoff

Current Advances

Nanofertilizers are fertilizers that use nanoparticles or nanotechnology to improve nutrient efficiency and plant nutrition. These have been shown to have positive effects on crop growth, yield, quality, and stress tolerance in various studies conducted at both national and international levels. Some of the success stories of nanofertilizers are given as follows:

- Nanofertilizers can increase the NUE of crops, which is the ratio of nitrogen uptake by plants to nitrogen input by fertilizers. A study using maize as a model system found that NUE from nanofertilizer was 82%, while NUE from conventional fertilizer (urea) was 42%, indicating a net higher NUE of 40% (Fatima et al., 2021). This suggests that nanofertilizers can provide a smart and controlled release of nutrients that match the crop requirement. Similar improvements in NUE were observed for other nutrients, such as phosphorus, zinc, iron, and magnesium, after applying nanoparticles (Avila-Quezada et al., 2022). Nanophosphorus also helps in increasing the organic acid concentration in the rhizosphere and phosphorus uptake by plants
- Nanofertilizers can mitigate the effects of abiotic stresses, such as drought, heat, salinity, waterlogging, and cold, that cause significant crop losses worldwide. A study using tomato as a model system evaluated the role of nanocalcium (LITHOVIT®), glycinebetaine (GB), acetylsalicylic acid (aspirin), and monopotassium phosphate fertilizers in alleviating salt stress, and found that LITHOVIT was the most effective in increasing the yield by 76% (Bhardwaj et al., 2022). Another study found that nanosilicon was more effective and efficient than regular silicon in alleviating salinity stress in rice (Avila-Quezada et al., 2022). This could be due to the changes in the physicochemical properties of the particles when they are modified into nanoformulation, such as higher surface area to volume ratio

Conclusion

Nanofertilizers are a promising alternative to conventional fertilizers, as they can provide nutrients in optimal amounts and in smaller quantities, enhance soil fertility and crop productivity, reduce environmental pollution and cost of fertilization, and improve crop quality and resilience. However, nanofertilizers also have some challenges and limitations that need to be addressed, such as the safety of nanomaterials for human health and the environment, the regulation and standardization of nanofertilizers, the accessibility and affordability of nanofertilizers for farmers, and the gap between research and application of nanofertilizers. Therefore, more research and development is needed to overcome these challenges and to make nanofertilizers more safe, reliable, and widely available. Nanofertilizers are a new and emerging technology that requires more education and training for farmers, consumers, and regulators. Nanofertilizers have a great potential for sustainable agriculture, but they also require further improvement and evaluation.

References

Avila-Quezada, G.D., Ingle, A.P., Golińska, P., Rai, M. "Strategic applications of nano-fertilizers for sustainable agriculture: Benefits and bottlenecks." *Nanotechnology Reviews* 11, 1 (2022): 2123-40.

Bhardwaj, A.K., Arya, G., Kumar, R., Hamed, L., Pirasteh-Anosheh, H., Jasrotia, P., et al. "Switching to nanonutrients for sustaining agroecosystems and environment: the challenges and benefits in moving up from ionic to particle feeding." *Journal of Nanobiotechnology* 20, 1 (2022): 19.

Bose, P. 2020. The effect of nano-fertilizers on sustainable crop development. https://www.azonano.com/article.aspx?ArticleID=5614 [Last accessed February, 2024].

Butt, B.Z., Naseer, I. 2020. "Nanofertilizers." *Nanoagronomy.* United States: Springer International Publishing. pp. 125-52.

Fatima, F., Hashim, A., Anees, S. "Efficacy of nanoparticles as nanofertilizer production: a review." *Environmental Science and Pollution Research* 28, 2 (2021): 1292-303.

Production F., Godoy, D., Dewbre, J., Pin Amegnaglo, C.J., Soglo, Y.Y., Akpa, A.F., Bickel, M., et al. 2017. "The Future of Food and Agriculture: Trends and Challenges." *The Future of Food and Agriculture: Trends and Challenges*, volume 4. Rome: Food and Agriculture Organization of the United Nations; 2017.

Iqbal, M., Umar, S., Mahmooduzzafar. "Nano-fertilization to Enhance Nutrient Use Efficiency and Productivity of Crop Plants." *Nanomaterials and Plant Potential*. Edited by Husen, A., Iqbal, M. Germany: Springer International Publishing. pp. 473-505.

KhetiGaadi. 2022. What are Nano Fertilizers? https://khetigaadi.com/blog/what-are-nano-fertilizers/ [Last accessed February, 2024].

Kumar, Y., Tiwari, K., Singh, T., Sain, N., Laxmi, S., Verma, R., et al. Nanofertilizers for enhancing nutrient use efficiency, crop productivity and economic returns in winter season crops of Rajasthan. Annals of Plant and Soil Research. 22, 4 (2020): 324-35.

Mohan, S.S., Regatti, V., Arigela, A. 2022. "Applications of Nano Fertilizers in Indian Agriculture." *Agriculture & Food E-Newsletter* 4, 5 (2022): 341-3.

Nongbet, A., Mishra, A.K., Mohanta, Y.K., Mahanta, S., Ray, M.K., Khan, M., et al. "Nanofertilizers: A Smart and Sustainable Attribute to Modern Agriculture." *Plants* 11, 19 (2022): 2587.

Pati, S., Dash, A., Maity, S., Pattnaik, S., Mohapatra, S., Samantaray, D.P. 2021. "Nano Biofertilizer for Sustainable Agriculture." *Agriculturally Important Microorganisms: Mechanisms and Applications for Sustainable Agriculture*. Edited by Pahari A., Mishra B.B., Nayak S.K. India: New India Publishing Agency. pp. 201-8.

Pramanik, P., Krishnan, P., Maity, A., Mridha, N., Mukherjee, A., Rai, V. 2020. "Application of Nanotechnology in Agriculture." *Environmental Nanotechnology*, volume 4. Germany: Springer International Publishing. pp. 317-48.

Rajput, V.D., Minkina, T., Feizi, M., Kumari, A., Khan, M., Mandzhieva, S., et al. "Effects of silicon and silicon-based nanoparticles on rhizosphere microbiome, plant stress and growth." *Biology* 10, 8 (2021): 791.

Sambangi, P., Gopalakrishnan, S., Pebam, M., Rengan, A. K. "Nano-biofertilizers on soil health, chemistry, and microbial community: benefits and risks. *Proceedings of the Indian National Science Academy* 88, 3 (2022): 357-68.

Suppan, S. 2017. Applying Nanotechnology to Fertilizer. Rationales, research, risks and regulatory challenges. https://www.iatp.org/documents/applying-nanotechnology-fertilizer-rationales-research-risks-and-regulatory-challenges [Last accessed February, 2024].

Tiwari, S.K. 2022. Nano-Fertilizers: A New Way to Enhance Nutrient Use Efficiency & Crop Productivity. https://krishijagran.com/agripedia/nano-fertilizers-a-new-way-to-enhance-nutrient-use-efficiency-crop-productivity/ [Last accessed February, 2024].

Verma, K.K., Song, X.P., Joshi, A., Rajput, V.D., Singh, M., Sharma, A., et al. "Nanofertilizer Possibilities for Healthy Soil, Water, and Food in Future: An Overview." *Frontiers in Plant Science* 13 (2022).

Verma, K.K., Song, X.P., Joshi, A., Tian, D.D., Rajput, V.D., Singh, M., et al. "Recent Trends in Nano-Fertilizers for Sustainable Agriculture under Climate Change for Global Food Security." *Nanomaterials* 12, 1 (2022): 173.

14

Nanopesticides

Abstract

Pesticides are chemicals that are used to control pests and diseases that affect crops and food production. However, pesticides have many problems, such as environmental contamination, bioaccumulation, pest resistance, and toxicity. Therefore, there is a need to reduce the amount and improve the safety of pesticides. Nanotechnology is a new and promising technology that can help achieve this goal. Nanotechnology is the science and engineering of manipulating matter at the nanoscale, which is about 1–200 nanometers (nm). Nanotechnology can create new or different properties in materials that are impossible to achieve by conventional methods. Nanotechnology can also improve the quality and shelf life of food products through smart packaging. One of the applications of nanotechnology in agriculture is the development of nanopesticides, which are nanostructures that contain or deliver pesticide-active ingredients. Nanopesticides have many advantages over conventional pesticides, which are mentioned as follows:

- They increase the solubility and stability of active ingredients, which means they can dissolve and remain effective better
- They release active ingredients slowly and steadily, which means they can provide a longer and more controlled protection
- They improve the mobility and surface area of active ingredients, which means they can reach and cover the target sites better
- They reduce the amount and wastage of active ingredients, which means they can save cost and resources
- They provide an eco-friendly approach, which means they can reduce the environmental impact and toxicity of pesticides

Keywords: Nanopesticide, bioaccumulation, nanotechnology, active ingredient, toxicological risk

Introduction

Nanopesticides are pesticides that use nanomaterials or nanotechnology to improve their effectiveness and safety in agriculture. Nanomaterials are materials that have at least one dimension between 1 and 200 nm. Nanomaterials have different properties than their bulk counterparts, such as higher surface area, reactivity, and mobility. Nanopesticides can be formulated in various ways, such as nanosized active ingredients, molecular active ingredient nanocarrier complexes, or active ingredients encapsulated or coated by nanomaterials (Deka et al., 2021; Xu et al., 2022). Nanopesticides can offer several benefits over conventional pesticides, which are mentioned as follows:

- They can increase the solubility, stability, and bioavailability of active ingredients, which means they can dissolve and remain effective better
- They can release active ingredients slowly and steadily, which means they can provide a longer and more controlled protection
- They can improve the mobility and surface area of active ingredients, which means they can reach and cover the target sites better
- They can reduce the amount and wastage of active ingredients, which means they can save cost and resources
- They can provide an eco-friendly approach, which means they can reduce the environmental impact and toxicity of pesticides

However, nanopesticides also have some challenges and limitations that need to be addressed. These are listed as follows:

- They may pose potential risks to human health and the environment due to their novel properties and interactions with biological systems (Chaud et al., 2021; Deka et al., 2021; Xu et al., 2022). Nanopesticides may accumulate in plant parts, soil, water, or food chains, and cause adverse effects such as oxidative stress, inflammation, or toxicity
- They are not well-regulated by the existing laws and standards for conventional pesticides or nanomaterials (Deka et al., 2021; Pan et al., 2023). There is a lack of clear definitions, classifications, guidelines, and protocols for nanopesticides
- They are still new and emerging technology that requires more research and development to evaluate their efficacy, safety, and environmental impact under field conditions and across different crops and regions (Deka et al., 2021; Pan et al., 2023; Xu et al., 2022). There is also a need for more education and training for farmers, consumers, and regulators

Nanopesticides are a promising alternative to conventional pesticides, but they also require further improvement and assessment. Nanopesticides have a great potential for sustainable agriculture, but they also require more caution and responsibility.

Motive and Rationale

Nanotechnology is the application of nanomaterials or nanoscale techniques to various fields, such as food, medicine, and agriculture. Nanomaterials are materials that have at least one dimension between 1 and 200 nm, which is about one billionth of a meter. Nanomaterials have different properties than their larger counterparts, such as higher surface area, reactivity, and mobility. Nanotechnology can offer several benefits for agriculture. These are:

- Improving crop productivity and quality by enhancing nutrient efficiency, soil fertility, plant growth, and disease resistance (Gangwar et al., 2023; Mittal et al., 2020; Nile et al., 2020)
- Reducing the use and wastage of fertilizers, pesticides, herbicides, and other agrochemicals by providing smart and controlled delivery systems
- Reducing the environmental impact and toxicity of agrochemicals by using eco-friendly and biodegradable nanomaterials
- Improving the detection and monitoring of pests, diseases, nutrients, and contaminants by using nanosensors and nanobiosensors
- Improving the quality and shelf life of food products by using nanomaterials for food packaging, preservation, and processing (Ahmad et al., 2022)

However, nanotechnology also has some challenges and limitations that need to be addressed.

- Assessing the safety and risk of nanomaterials for human health and the environment by conducting more research and testing under field conditions and across different crops and regions (Gangwar et al., 2023; Mittal et al., 2020; Nile et al., 2020)
- Developing and implementing appropriate regulations and standards for nanotechnology in agriculture by involving different stakeholders and agencies (Ahmad et al., 2022; Nile et al., 2020)
- Enhancing the accessibility and affordability of nanotechnology for farmers by reducing the cost and complexity of nanomaterials and nano-based technologies (Ahmad et al., 2022; Nile et al., 2020)

- Enhancing the education and awareness of nanotechnology for consumers and regulators by providing more information and training on the benefits and challenges of nanotechnology in agriculture (Ahmad et al., 2022; Nile et al., 2020)

Formulations

Nanopesticides are pesticides that use nanomaterials or nanotechnology to improve their effectiveness and safety in agriculture. As shown in Figure 1, nanopesticides can be formulated in various ways, depending on their intended purpose and composition (Chaud et al., 2021; Pan et al., 2023). Some of the common types of nanopesticide formulations are:

- *Nanoemulsions*: These are oil-in-water (O/W) emulsions, where the active ingredient of the pesticide is dispersed as nanosized droplets in water, with surfactant molecules confined at the pesticide–water interface. Nanoemulsions can be thermodynamically stable or kinetically stable, depending on the solubility of the active ingredient and the amount and type of surfactants. Nanoemulsions can improve the solubility, stability, and dispersion of active ingredients
- *Nanosuspensions*: These are suspensions of solid nanosized particles of active ingredients in aqueous media, with surfactant molecules confined at the particle surface. Nanosuspensions can increase the surface area, reactivity, and mobility of active ingredients
- *Polymer-based nanoparticles*: These are nanoparticles made of polymers or copolymers that can carry or deliver active ingredients to the target sites. Polymer-based nanoparticles can be classified into different subtypes, such as nanocapsules, nanospheres, nanogels, and nanofibers
- *Nanocapsules*: These are heterogeneous reservoir-type structures, where the active ingredient is enclosed within a polymer coating or membrane. Nanocapsules can protect the active ingredient from degradation and control its release
- *Nanospheres*: These are homogeneous vesicular structures, where the active ingredient is uniformly dispersed throughout the polymer matrix. Nanospheres can enhance the stability and dispersion of active ingredients
- *Nanogels*: These are hydrogel nanoparticles, where the polymer matrix contains hydrophilic groups that can absorb water. Nanogels can improve the solubility and delivery of active ingredients

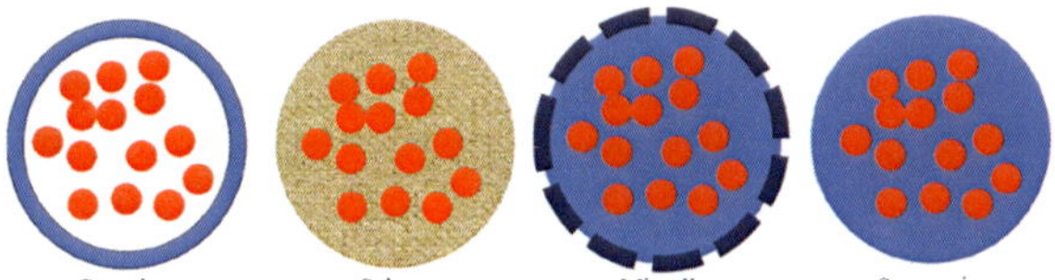

Fig. 1: Different compositions of nanopesticides.

- *Nanofibers*: These are fibrous structures made of polymers or copolymers that can load or release active ingredients. Nanofibers can be produced by electrospinning, thermal-induced phase separation, or other methods.

Modes of Loading

Nanoencapsulation is a technique of enclosing one biologically active ingredient within a nanoscale coating or matrix, with the aim of controlling the release and delivery of the ingredient to the target site. Nanoencapsulation can protect the ingredient from degradation, evaporation, leaching, or drift, and enhance its solubility, stability, mobility, and efficacy. Nanoencapsulation can also reduce the amount and toxicity of the ingredient, and provide an eco-friendly approach for agriculture. Nanoencapsulation can be applied to various types of pesticides, such as insecticides, herbicides, fungicides, and nematicides (Hajji-Hedfi and Chhipa 2021; Madhavi et al., 2020).

- *Carbon nanotubes*: These are cylindrical structures made of carbon atoms that can load or deliver pesticides. Carbon nanotubes have shown positive effects on plant growth, biomass, root length, and seed germination. However, carbon nanotubes may also have negative effects on plant cells and soil microbes due to their reactive oxygen species (ROS) generation
- *Polyelectrolyte complexes*: These are complexes formed by electrostatic interactions between oppositely charged polymers that can load or deliver pesticides. Polyelectrolyte complexes can reduce the dose and leaching rate of pesticides. Polyelectrolyte complexes can also be used as biosensors to detect food poisoning
- *Clay and layered double hydroxides (LDHs)*: These are mineral-based nanoparticles that can bind or deliver pesticides. Clay and LDHs can protect pesticides from volatilization and photodegradation. Clay and LDHs are considered safer for the ecosystem due to their reduced drift and leaching rates

- *Organically modified silica nanoparticles*: These are nanoparticles made of silica that are modified with organic groups that can load or deliver pesticides. Organically modified silica nanoparticles have the versatility of organic polymers and other advantages due to the sol–gel process, such as tailoring shape, density, and surface property. Organically modified silica nanoparticles can provide a high load of pesticide that can be released over a long time and remain stable in harsh environmental conditions

Nanoencapsulated pesticides have many advantages over conventional pesticides, but they also have some challenges and limitations that need to be addressed. Some of these challenges are assessing their safety and risk for human health and the environment (Hajji-Hedfi and Chhipa, 2021; Madhavi et al., 2020; Tao et al., 2023), developing and implementing appropriate regulations and standards for their use, enhancing their accessibility and affordability for farmers, enhancing their education and awareness for consumers and regulators, and comparing their efficacy and impact with existing commercial products under field conditions (Hajji-Hedfi and Chhipa 2021; Tao et al., 2023). Nanoencapsulated pesticides are a promising alternative to conventional pesticides, but they also require more caution and responsibility. Nanoencapsulated pesticides have a great potential for sustainable agriculture, but they also require more innovation and evaluation.

Current Advances

Nanopesticides have been shown to have positive effects on crop protection against various pests, diseases, and weeds. Some examples of nanopesticide research and development are:

- Beta-cyfluthrin nanoformulation showed prolonged activity and higher effectiveness against *Callosobruchus maculatus*, a pest of stored legumes (Loha et al., 2011)
- Nanoacephate prepared by encapsulation of acephate in polyethylene glycol (PEG) was more promising against *Spodoptera litura*, a pest of cotton and other crops. It did not induce any cytotoxicity in human fibroblast cell line (Pradhan et al., 2013)
- A nanogel prepared from a pheromone, methyl eugenol, using a low-molecular-mass gelator was very stable at open ambient conditions and slowed down the evaporation of pheromone significantly. It enabled its easy handling and transportation without refrigeration and reduction in the frequency of pheromone recharging in the orchard. It also brought

about an effective management of oriental fruit fly *Bactrocera dorsalis*, a harmful pest for many fruits including guava (Bhagat et al., 2013)

- Silica nanoparticles showed potential as a new insecticide for pest control. They enhanced the activity and stability of mancozeb, an antifungal pesticide. They also increased water solubility of active ingredients. However, they also caused negative outcomes on plants and soil microbes due to ROS formation (El-Bendary and El-Helaly 2013)
- Polymer stabilized bifenthrin nanoparticles were developed as nanospheres. They showed higher efficacy and lower toxicity than conventional bifenthrin against *Aedes aegypti* mosquitoes (Liu et al., 2008)
- Neem-based nanocapsules prepared using biodegradable polymers, such as poly-ε-caprolactone (PCL) and poly-β-hydroxybutyrate (PHB), caused higher mortality of *Bemisia tabaci*, a serious pest of many crops (de Carvalho et al., 2015)
- PEG nanoparticles with essential oil extracted from *Geranium* sp. and *Citrus reticulata* showed slow release of terpenes that enhanced the toxicity multifold to insect *Blattella germanica* (Werdin González et al., 2015)
- Nanoemulsion containing apolar fraction from fruits of *Manilkara subsericea* exhibited significantly higher levels of mortality of cotton pest *Dysdercus peruvianus* (Fernandes et al., 2014)
- Similarly nanoemulsion developed using *Simmondsia chinensis* (Jojoba) seed oil and Tween-20 as a nonionic surfactant and water was more effective against *Sitophilus oryzae*, a pest of stored grains, than the bulk jojoba oil (Sh et al., 2015)

Conclusion

Nanotechnology is a novel and emerging technology that can offer many benefits for agriculture and food production. One of the applications of nanotechnology is the development of nanopesticides, which are nanostructures that contain or deliver pesticide active ingredients. Nanopesticides can enhance the efficiency and effectiveness of pesticides, while reducing the amount and toxicity of pesticides. Nanopesticides can also provide a more sustainable and eco-friendly approach for pest and disease control. Therefore, nanotechnology can be a potential solution for the problems of conventional pesticides and can improve the quality and safety of food products. However, nanotechnology also poses some challenges and risks, such as the lack of regulation, standardization, and safety assessment. Therefore, more research and development are needed to explore the potential and limitations of nanotechnology in agriculture and food production.

References

Ahmad, Z., Tahseen, S., Wasi, A., Ganie, I.B., Shahzad, A., Emamverdian, A., et al. "Nanotechnological Interventions in Agriculture." *Nanomaterials* 12, 15 (2022): 2667.

Bhagat, D., Samanta, S.K., Bhattacharya, S. "Efficient Management of Fruit Pests by Pheromone Nanogels." *Scientific Reports* 3, 1 (2013): 1294.

de Carvalho, S.S., Vendramim, J.D., Sá, I.C.G. da Silva, M.F. das G.F., Ribeiro, L. do P., et al. "Systemic insecticidal effect of neem-based nanoformulations against Bemisia tabaci (Hemiptera: Aleyrodidae) biotype B in tomato." *Bragantia* 74 (2015): 298-306.

Chaud, M., Souto, E.B., Zielinska, A., Severino, P., Batain, F., Oliveira-Junior, J., et al. "Nanopesticides in Agriculture: Benefits and Challenge in Agricultural Productivity, Toxicological Risks to Human Health and Environment." *Toxics* 9, 6 (2021): 131.

Deka, B., Babu, A., Baruah, C., Barthakur, M. "Nanopesticides: A Systematic Review of Their Prospects With Special Reference to Tea Pest Management." *Frontiers in Nutrition* 8 (2021).

El-Bendary, H.M., El-Helaly, A.A. "First record nanotechnology in agricultural: Silica nano-particles a potential new insecticide for pest control." *Applied Science Reports* 4, 3 (2013): 241-6.

Fernandes, C.P., de Almeida, F.B., Silveira, A.N., Gonzalez, M.S., Mello, C.B., Feder, D., et al. "Development of an insecticidal nanoemulsion with Manilkara subsericea (Sapotaceae) extract." *Journal of Nanobiotechnology* 12 (2014): 1-9.

Gangwar, J., Kadanthottu Sebastian, J., Puthukulangara Jaison, J., Kurian, J.T. "Nano-technological interventions in crop production—a review." *Physiology and Molecular Biology of Plants* 29, 1 (2023): 93-107.

Hajji-Hedfi, L., Chhipa, H. "Nano-based pesticides: challenges for pest and disease management." *Euro-Mediterranean Journal for Environmental Integration* 6, 3 (2021): 69.

Liu, Y., Tong, Z., Prud'homme, R.K. "Stabilized polymeric nanoparticles for controlled and efficient release of bifenthrin." *Pest Management Science* 64, 8 (2008): 808-12.

Loha, K., Shakil, N., Kumar, J., Singh, M., Adak, T., Jain, S. "Release kinetics of β-Cyfluthrin from its encapsulated formulations in water." *Journal of Environmental Science and Health, Part B* 46, 3 (2011): 201-6.

Madhavi, V., Reddy, A.V.B., Madhavi, G., Reddy, N.B. "Nanoencapsulation of pesticides: Sustainable perspective in agriculture." *AIP Conference Proceedings* (2020): 2280.

Mittal, D., Kaur, G., Singh, P., Yadav, K., Ali, S.A. "Nanoparticle-Based Sustainable Agriculture and Food Science: Recent Advances and Future Outlook." *Frontiers in Nanotechnology* 2 (2020).

Nile, S.H., Baskar, V., Selvaraj, D., Nile, A., Xiao, J., Kai, G. "Nanotechnologies in Food Science: Applications, Recent Trends, and Future Perspectives." *Nano-Micro Letters*, 12, 1 (2020): 45.

Pan, X., Guo, X., Zhai, T., Zhang, D., Rao, W., Cao, F., et al. "Nanobiopesticides in sustainable agriculture: developments, challenges, and perspectives." *Environmental Science: Nano* 10, 1 (2023): 41-61.

Pradhan, S., Roy, I., Lodh, G., Patra, P., Choudhury, S.R., Samanta, A., et al. "Entomotoxicity and biosafety assessment of PEGylated acephate nanoparticles: A biologically safe alternative to neurotoxic pesticides." *Journal of Environmental Science and Health Part B* 48, 7 (2013): 559-69.

Sh, A., Abdelrazeik, A.B., Rakha, O.M. "Nanoemulsion of jojoba oil, preparation, characterization and insecticidal activity against Sitophilus oryzae (coleoptera: Curculionidae) on wheat." *International Journal of Agriculture Innovations and Research* 4 (2015): 72-5.

Tao, R., You, C., Qu, Q., Zhang, X., Deng, Y., Ma, W., et al. "Recent advances in the design of controlled- and sustained-release micro/nanocarriers of pesticide." *Environmental Science: Nano* 10, 2 (2023): 351-71.

Werdin González, J.O., Stefanazzi, N., Murray, A.P., Ferrero, A.A., Fernández Band, B. "Novel nanoinsecticides based on essential oils to control the German cockroach." *Journal of Pest Science 88* (2015): 393-404.

Xu, Z., Tang, T., Lin, Q., Yu, J., Zhang, C., Zhao, X., et al. Environmental risks and the potential benefits of nanopesticides: a review. *Environmental Chemistry Letters* 20, 3 (2022): 2097-108.

15

Simulation Models

Abstract

A simulation model that explains how crop growth and development are influenced by soil properties, weather, and crop management is called a crop simulation model or CSM. These models usually predict when particular growth stages will be reached, the biomass of crop components (leaves, stems, roots, and harvestable products) as they vary over time, and corresponding variations in soil moisture and nutrient status. These dynamic models aim to replicate crop growth and development by utilizing basic mechanisms found in plant and soil processes. Although the algorithms are not all the same, most of them have a time step of one day. These are the main instruments at our disposal to evaluate the possible effects of climate change on crop productivity and to enhance our comprehension of crop performance, yield gaps, genetic gains, better prediction of pest and insect outbreaks, more efficient irrigation systems, and optimized planting dates.

Keywords: Crop simulation models, climate change, yield gap, dynamic models

Introduction

Crop models can be broadly classified into two categories: dynamic or process-based crop simulation models (CSMs), and reduced form statistical models. Since simulation models are based on various plant-related processes, they develop or alter in tandem with plant growth (Chetty 2009); as a result, they are inherently nonlinear (Olmstead 2009). To run, calibrate, and validate, they require a significant amount of date input. However, statistical models are helpful in situations where there is a lack of data because they describe the relationships between a subset of variables while keeping the others constant. (Chetty 2009).

History

The first rice models were created in the 1970s and 1980s, marking the beginning of crop modeling history. Subsequently, CGIAR centers ORYZA on rice, WARDA on rice, ICRISAT on groundnut, CIT on beans, and CIMMYT on maize developed crop models starting in the 1990s. ICRISAT (on soybean, groundnut, and sorghum), CIP (on potatoes), IRRI (rice model), CIAT (on dry beans and legumes), CIMMYT (on maize breeding, crop-livestock interactions), ILRI (on crop-livestock interactions), and International Food Policy Research Institute (IFPRI) (Impact of climate change on agriculture, soil monitoring) developed innovative crop models at CGIAR centers during 2000. In 2010, there was a surge in interest in regional and global modeling, with a particular focus on climate change foresight work, strong connections with AgMIP, UF modeling groups, and CSIRO modeling activities encompassing agronomy, rotations, and agroforestry, among other applications. According to Reynolds (2018), these centers included CIAT, IFPRI, IRRI, CIMMYT, ICRISAT, CIP, ICRAF, ILRI, IITA, and ICARDA. The Decision Support System for Agrotechnology Transfer (DSSAT) model (Hoogenboom et al., 2012; Jones et al., 2003), the Agricultural Production Systems Simulator Model (APSIM) (Keating et al., 2003), and the Global Agro-Ecological Zone (AEZ) modeling framework (Fischer et al., 2002; Fischer et al., 2005) are the most widely used simulation models for the majority of crop simulation work. Along with DSSAT, EPIC, DSSAT, PEGASUS (Nelson et al., 2014), and the General Large Area Model (GLAM) for annual crops (Challinor et al., 2010), the Lund-Potsdam-Jena managed Land (LPJmL) model (Bondeau et al., 2007) has also been used in more recent work (Blanco et al., 2014; Frank et al., 2014).

Fundamental Idea, Steps, and Materials Needed for Crop Simulation Modeling

The processes of crop growth and development are described by a CSM, which is a dynamic model. These processes are based on the genetic potential of the cultivar (G), the physical environmental conditions (E) of the crop, and crop management inputs and their interactions with one another, or G × EÏM. These three elements are also indirectly impacted by socioeconomic conditions (S), which include the supply and demand for inputs and outputs, the amount of credit required and taken, agricultural policies, and the practices adopted by farmers and consumers. These factors extend the paradigm to GÏEÏMΏS. Genetic coefficients of cultivars are computed in order to get G. For that reason, a field experiment or controlled experiment is set up to track the crop's phenological growth every day, from the time of germination to harvesting. For E, soil data (depths of the major soil horizons, particle size analysis, BD, water release curve and saturated HC, residual fertilizer content at the beginning of the season, organic matter content at planting, and soil temperature) and detailed weather (T_{max} and T_{min}, total solar radiation, precipitation, humidity

(09:00 hours, and wind speed) are taken. The latitude of the site, row spacing, plant population, amount and timing of fertilizer applications, and other pertinent information are examples of management data. The supply and demand of inputs and outputs, credit and financing, agricultural policies, and the adaptable farming and consumer practices are all taken into account for S.

To run the CSM, all of these input data are required. To examine the relationship between experimental data and data produced by the model, model calibration or verification must be carried out after the model has been run. Many times, the simulated values differ slightly from the observed data, necessitating small adjustments for those parameters in order to keep the model reasonably close to the actual values of the experimental trials. We call this process calibration. In order to explain the model's dependability and fully examine the relationship between a model's predictions and its input parameters, sensitivity analysis is carried out.

Following the necessary validations and calibrations, the data is simulated to forecast plant growth and development in relation to crop management and environmental factors, which are supplied as input data to the model (Hodson and White 2010).

Crop Models and Futuristic Approaches for

Food Security

Crop models have become more and more popular in recent years as a means of analyzing potential effects of climate change on crop growth and global food production. Based on this methodology, studies have demonstrated that by 2050, the effects of climate change could lower global yields of wheat, rice, and corn by as much as 25% when compared to a baseline where there is no climate change and before economic adjustments are taken into account (Rosegrant et al., 2014). In order to obtain simulated yields under various climate scenarios and production environments, the IFPRI has made a significant and ongoing effort to investigate changes in the productivity of major food crops worldwide. The results obtained are organized, exported, and run through crop models. The output data obtained is then organized, manipulated, analyzed, and finally interpreted to provide context as well as specific information so policymakers can plan appropriately estimate the future requirement of foodgrains for the worldwide population (Robertson 2017).

Climate Change Studies

Numerous CSMs have been created and tested in order to assess the possible effects of climate change on agricultural output. These models provide information that helps mitigate risks and comprehend the workings of physical and biological systems. Additionally, by identifying the traits that have the biggest effects on yield and the futuristic crop management techniques, they can support crop improvement programs. However, there are a number of uncertainties in the model-based assessment of climate change that can be minimized by combining multimodeling with crop modeling that incorporates plant physiology, biochemistry, and gene-based modeling. To do this, it is necessary to carry out precise field experiments and produce high-quality field data for model testing.

Fusion with Intelligent Farming

The role that geographic information systems (GIS), Global Positioning Systems (GPS), and intelligent implements play is another area where technology is developing quickly. These methods can be combined with simulation models to improve decision-making. By combining these smart technologies with crop models, predictions can be made that combine the spatial perspective of GIS with the more robust temporal process representation of simulation models. Maps are particularly useful for helping people comprehend how different regions are affected by climate change and how different adaptations might be made. (Hodson and White 2010). Model results can be used to control harvest timing, nitrogen fertilizer, irrigation water, density of the population, row spacing, and the application of plant growth regulators (Boone et al., 1997). Crop models and precision agriculture could help policymakers and breeders to make further decisions for sustainable growth.

Adoption of New Technologies via Crop Models

Crop models aid in the adoption of new technologies even more. In practice, an extreme climate scenario was simulated using crop models in order to test the benefits of new technologies in comparison to testing new varieties under more extreme climate scenarios. The drought and heat-resistant varieties that were found to be appropriate for this scenario performed well and were predicted to yield greater benefits; as a result, they were recommended for further use in the future under climate change scenarios (Islam et al., 2016). When incorporated into crop models, survey samples further elucidate the adoption of new resource-saving technologies by boosting exposure and awareness through planned field days and demonstration trials. In addition, free access to expensive machinery for first-time users increases the likelihood, tempo, and intensity of adoption (Yigezu et al., 2018).

Molecular Science and Crop Models

In the future, molecular geneticists could potentially utilize CSMs as a tool to identify the genetic traits that would yield the highest yields and returns under specific environmental conditions. They also take into account the pleiotropic effects of changing traits. For example, while increasing protein in a crop is a desirable trait, CSMs clearly show that yield decreases by approximately 1% for every 1% increase in protein. In addition, scientists consider numerous desirable traits at the molecular level, but the models clearly assist the scientists in accepting the challenges of thinking about these character modifications more precisely (Boote et al., 1996)

Pest Forecasting

Worldwide crop production is threatened by pests and diseases, which result in losses of 10%–28% for wheat, 25%–41% for rice, 20%–41% for maize, 8%–21% for potatoes, and 11%–32% for soybeans (Savary et al., 2019). These random organisms undergo mutations that result in the production of new pathotypes that are unpredictable and hard to control (Reynolds 2018). Thus, it is difficult to comprehend, control, and predict how pests will affect crop production (Garrett et al., 2013). Insect pests and their interactions with plants, weather, nitrogen, water control, supply, demand, and other factors are the main subjects of modeling efforts (Ittersum et al., 2003). In pest management, modeling not only aids in pest population forecasting, but it also provides solutions for issues like state selection, strategy selection, and tactic selection (Dalal and Singh 2017). Similar to this, the success of pest control programs depends on focusing on the stage of the pest's life cycle that will destroy the plant the most effectively (Diuk-Wasser et al., 2010). In this case, using a model to simulate phonological stages could yield information about the percentage of each life stage in the species (Sporleder et al., 2013).

Agro Advisory

Agro advisory refers to agricultural decisions made in response to changing weather conditions, primarily during flood and drought, including agronomical, pest and disease, water, and input management. Farmers, businesses reliant on agriculture, producers of animal feed, the government, and others need it to make policy decisions and market agricultural products. This Agromet Advisory was issued in response to changes in the weather in the past, present, and future. It maintains crop productivity while boosting input effectiveness.

Farmers now have a better understanding of farming techniques, including information on cutting-edge technologies and recommended practices, thanks to agro advisories (Chhabra and Kumar 2018). When unusual conditions, such as drought, occur in a region, agro advisories are issued based on 30- to 40-year weather data patterns, with models generating predictions based on that variability. Researchers can determine intervention strategies and points of intervention in the farm value chain that yield the highest returns for the least amount of risk by utilizing system modeling tools in conjunction with seasonal and subseasonal climate forecasts (CGIAR, 2018).

Constraints of Crop Models

The agricultural system is recognized for its significant interplay between its incomprehensible biotic and abiotic constituents. Thus, models only depict a rough version of reality. When knowledge is lacking, the modeler typically uses a simplified equation to explain a complex subsystem. As a result, simplifications are made with the intention of modeling knowledge and are thus biased. Models that are the product of weak interdisciplinary collaboration are frequently weak in other areas but strong in those where the developers are experts.

The caliber of scientific data used in the development, calibration, and validation of the model also affects its quality. When introducing the crop model into a new situation, the calibration and validation steps are critical to its success. The quality of the input data used determines how accurate the results are. In these situations, using approximations could result in incorrect interpretations. Updated crop models rely on comprehensive meteorological data. Again, the model's results become erroneous if vital parameters like solar radiation are missing from the data (Kumar and Chaturvedi 2009).

In order to overcome all of these constraints the model needs to account for the variability of the data that it is supposed to be fed (Wisiol 1987). A model needs to be as sensitive as the system it represents, meaning it should be able to detect variations in the input parameters and generate disparate outcomes as the real system would (Dourado et al., 1998).

Conclusion

A potentially invaluable tool for researchers, crop management professionals, and policy makers is crop growth modeling. For it to function, inputs like crop and weather data are needed. In order to assess the possible effects of climate change on crop productivity and to inform agricultural adaptation strategies for risk management and increased food grain security in the face of changing climate conditions, CSMs can be very helpful. With its integration with smart agriculture and ability to manage diseases and pests, it can be a valuable tool for policy makers making decisions about the adoption of new technologies and ensuring food security for the world's expanding population. Nonetheless, there is a pressing need to find efficient ways to get the results of work employing models out to potential beneficiaries and to make the use of models in research more relevant to solving real-world problems. Notwithstanding certain drawbacks, the modeling technique continues to be the most effective way to evaluate the consequences of potential future global climate change, which aids in the development of national policies for climate-resilient and increased food security.

References

FAO, IFAD, UNICEF, WFP, WHO). 2019. *The State of Food Security and Nutrition in the World. Safeguarding against Economic Slowdowns and Downturns*. Rome: FAO.

Blanco, M., Ramos, F., Van Doorslaer, B., 2014. "Economic impacts of climate change on agrifood markets: A Bio-Economic approach with a focus on the EU." *Proceedings of the XIVth EAAE Congress*. Ljubbljana: Agri-Food and Rural Innovations for Healthier Societies.

Bondeau, A., Smith, P.C., Zaehle, S., Schaphoff, S., Lucht, W., Cramer, W., et al. Modelling the role of agriculture for the 20th century global terrestrial carbon balance. *Global Change Biology* 13, 3 (2006): 679-706.

Boone M.Y.L., Kikusawa M., McKinion J.M. 1997. "Crop models and precision agriculture." *Applications of Systems Approaches at the Field Level.* Edited by Kropff, M.J., Teng, P.S., Aggarwal, P.K., Bouma, J., Bouman, B.A.M., Jones, J.W., et al. Dordrecht: Springer. pp. 189-99.

Boote, K.J., Jones, J.W., Pickering, B.N. "Potential Uses and Limitations of Crop Models." Agronomy Journal. 88 (1996): 704-16.

CGIAR. 2018. ILRI Vacancy: Scientist—Climate Information and Agro-Advisory. https://ccafs.cgiar.org/careers-and-calls/ilri-vacancy-scientist-climate-information-and-agro-advisory [Last accessed February, 2024].

Chakrabarty, M. 2016. Climate Change and Food Security in India. ORF Issue Brief. https://www.orfonline.org/wp-content/uploads/2016/09/ORF_IssueBrief_1571.pdf [Last accessed February, 2024].

Challinor, A.J., Simelton, E.S., Fraser, E.D., Hemming, D., Collins, M. "Increased crop failure due to climate change: assessing adaptation options using models and socio-economic data for wheat in China." *Environmental Research Letters* 5, 3 (2010): 034012.

Chetty, R. "Sufficient Statistics for Welfare Analysis: A Bridge Between Structural and Reduced-Form Methods." *Annual Review of Economics* 1, 1 (2009): 451-88.

Chhabra, K., Kumar, Y. "Agro-Advisory in Agriculture." *Marumegh* 3, 3 (2018).

Dalal, P.K. Singh, J.K. "Role of modeling in insect pest and disease management." *Journal of Entomology and Zoology Studies* 5, 5 (2017): 1773-7.

Diuk-Wasser M.A., Vourc'h G., Cislo P., Hoen A.G., Melton F., Hamer S.A., et al. "Field and climate-based model for predicting the density of host-seeking nymphal Ixodes scapularis, an important vector of tick-borne disease agents in the eastern United States." *Global Ecology and Biogeography* 19, 4 (2010): 504-14.

Dourado-Neto, D., Teruel, D.A., Reichardt, K., Nielsen, D.R., Frizzone, J.A., Bacchi, O.O.S., et al. "Principles of crop modeling and simulation: I. uses of mathematical models in agricultural science." *Scientia Agricola* 55 (1998).

Fischer, G., Shah, M., Van Velthuizen, H.T. *Climate change and agricultural vulnerability. Discussion Paper*. Laxenburg: IIASA.

Fischer, G., Shah, M., Tubiello, F.N., Van Velhuizen, H., "Socio-economic and climate change impacts on agriculture: an integrated assessment, 1990–2080." *Philosophical Transactions of the Royal Society of London Series B, Biological Sciences* 360, 1463 (2005): 2067-83.

Garrett, K.A., Dobson, A.D., Kroschel, M.J.B., Natarajan, S.O., Tonnang, H.E., Valdivia, C. "The effects of climate variability and the color of weather time series on agricultural diseases and pests, and on decisions for their management." Agric. For. Meteorol 170 (2013): 216-27.

Frank, S., Witzke, H.P., Zimmermann, A., Havlík, P., Ciaian, P. 2014. "August. climate change impacts on European agriculture: a multi model perspective." Proceedings of the XIVth EAAE Congress. Ljubbljana: Agri-Food and Rural Innovations for Healthier Societies.

Hodson, D. White, J. 2010. GIS and Crop Simulation Modelling Applications in Climate Change research https://www.researchgate.net/publication/48855991_GIS_and_Crop_Simulation_Modelling_Applications_in_Climate_Change_Research [Last accessed February, 2024].

Hoogenboom, G., Jones, J.W., Wilkens, P.W., Porter, C.H., Boote, K.J., Hunt, L.A., et al. 2012. *Decision Support System for Agrotechnology Transfer (DSSAT) Version 4*. Honolulu: University of Hawaii. p. 5.

Islam, S., Cenacchi, N., Sulser, T.B., Gbegbelegbe, S., Hareau, G., Kleinwechter, U., et al. "Structural approaches to modeling the impact of climate change and adaptation technologies on crop yields and food security." *Global Food Security* 10 (2016): 63-70.

Ittersum, M.K., Van, Leffelaar, P.A.H., Van, K., Kropff, M.J., Bastiaans, L., Goudriaan, J. "On approaches and applications of the Wageningen crop models." *European Journal of Agronomy* 18 (2003): 201-34.

Jones, J.W., Antle, J.M., Basso, B., Boote, K.J., Conant, R.T., Foster, I., et al. Brief history of agricultural systems modeling. *Agricultural Systems* 155, (2017): 240-54.

Kumar, D.P. Singh, J. "Role of modeling in insect pest and disease management." *Journal of Entomology and Zoology Studies* 5 (2017): 1773-7.

Kumar, M. "Impact of climate change on crop yield and role of model for achieving food security." *Environmental Monitoring and Assessment*. 188, 8 (2016): 465.

Kumar, R. Chaturvedi, S. 2009. Crop modeling: A tool for agricultural research http:// agropedia. Iitk .ac.in /content/crop-model.

Liu L, Basso B. "Impacts of climate variability and adaptation strategies on crop yields and soil organic carbon in the US Midwest." *PLoS One* 15, 1 (2020): e0225433.

Robertson, R. 2017. Mink: process-based crop modeling for global food security. https://globalfutures.cgiar.org/2017/09/05/mink-process-based-crop-modeling-for-global-food-security/ [Last accessed February, 2024].

Nelson, G.C., Valin, H., Sands, R.D., Havlík, P., Ahammad, H., Deryng, D., et al. "Climate change effects on agriculture: economic responses to biophysical shocks." *Proceedings of the National Academy of Sciences* 111, 9 (2014): 3274-9.

Reynolds, M., Kropff, M., Crossa, J., Koo, J., Kruseman, G., Milan, A.M., et al. "Role of Modelling in International Crop Research: Overview and Some Case Studies." *Agronomy* 8, 291 (2018).

Rosegrant, M.W., Koo, J., Cenacchi, N., Ringler, C., Robertson, R., Fisher, M., et al. 2014. *Food Security in a World of Natural Resource Scarcity: The Role of Agricultural Technologies*. Washington: IFPRI.

Savary, S., Willocquet, L., Pethybridge, S.J., Esker, P., McRoberts, N., Nelson, A. "The global burden of pathogens and pests on major food crops." *Nature Ecology and Evolution* 3, 3 (2019): 430-9.

Sporleder, M., Tonnang, H.E.Z., Carhuapoma, P., Gonzales, J.C., Juarez, H, Kroschel, J., et al. 2013. "Insect Life Cycle Modelling (ILCYM) software—a new tool for regional and global insect pest risk assessments under current and future climate change scenarios." *Potential Invasive Pests Agricultural Crops*. Edited by Peña, J.E. Wallingford: CABI. pp. 412-27.

Wisiol, K. 1987. "Choosing a basis for yield forecasts and estimates." *Plant Growth Modeling for Resource Management: Current Models and Methods*. Edited by Wisiol, K., Hesketh, J.D. Boca Raton: CRC Press. pp. 75-103.

Yigezu, A.Y., Mugera, A., Shater, T.E., Hassan, A.A., Piggen, C., Haddad, A., et al. "Enhancing adoption of agricultural technologies requiring high initial investment among smallholders." *Technological Forecasting and Social Change* 134 (2018): 199-206.

Index